BIOLOGY AND ECOLOGY
OF VENOMOUS STINGRAYS

BIOLOGY AND ECOLOGY OF VENOMOUS STINGRAYS

Ramasamy Santhanam, PhD

Apple Academic Press Inc.
3333 Mistwell Crescent
Oakville, ON L6L 0A2 Canada

Apple Academic Press Inc.
9 Spinnaker Way
Waretown, NJ 08758 USA

ISBN 13: 978-1-77-463668-8 (pbk)
ISBN 13: 978-1-77-188538-6 (hbk)

Library and Archives Canada Cataloguing in Publication

Santhanam, Ramasamy, 1946-, author
Biology and ecology of venomous stingrays / Ramasamy Santhanam, PhD.

(Biology and ecology of marine life)
Includes bibliographical references and index.
Issued in print and electronic formats.
ISBN 978-1-77188-538-6 (hardcover).--ISBN 978-1-315-20743-8 (PDF)
1. Stingrays. 2. Stingrays--Ecology. 3. Freshwater stingrays. 4. Freshwater stingrays--Ecology. 5. Rays (Fishes). 6. Rays (Fishes)--Ecology. I. Title. II. Series: Santhanam, Ramasamy, 1946-Biology and ecology of marine life

QL638.8.S26 2017	597.3'5	C2017-902627-5	C2017-902628-3

Library of Congress Cataloging-in-Publication Data

Names: Santhanam, Ramasamy, 1946-
Title: Biology and ecology of venomous stingrays / Ramasamy Santhanam, PhD.
Description: Toronto : Apple Academic Press, 2017. | Includes bibliographical references and index.
Identifiers: LCCN 2017017482 (print) | LCCN 2017018532 (ebook) | ISBN 9781315207438 (ebook) | ISBN 9781771885386 (hardcover : alk. paper)
Subjects: LCSH: Stingrays. | Stingrays--Ecology. | Freshwater stingrays. | Freshwater stingrays--Ecology. | Rays (Fishes) | Rays (Fishes)--Ecology.
Classification: LCC QL638.8 (ebook) | LCC QL638.8 .S26 2017 (print) | DDC 597.3/5--dc23
LC record available at https://lccn.loc.gov/2017017482

Apple Academic Press also publishes its books in a variety of electronic formats. Some content that appears in print may not be available in electronic format. For information about Apple Academic Press products, visit our website at **www.appleacademicpress.com** and the CRC Press website at **www.crc-press.com**

ABOUT THE AUTHOR

Ramasamy Santhanam, PhD

Dr. R. Santhanam is the former Dean of Fisheries College and Research Institute, Tamilnadu Veterinary and Animal Sciences University, Thoothukudi, India. His fields of specialization are marine biology and fisheries environment. Presently, he is serving as a resource person for various universities of India. He has also served as an expert for the Environment Management Capacity Building, a World Bank-aided project of the Department of Ocean Development, India. He was a Member of the American Fisheries Society, United States; World Aquaculture Society, United States; Global Fisheries Ecosystem Management Network (GFEMN), United States; and the International Union for Conservation of Nature's (IUCN) Commission on Ecosystem Management, Switzerland. To his credit, Dr. Santhanam has 21 books on fisheries science/marine biology and 70 research papers.

BIOLOGY AND ECOLOGY OF MARINE LIFE BOOK SERIES

Series Author:
Ramasamy Santhanam, PhD
Former Dean, Fisheries College and Research Institute,
Tamil Nadu Veterinary and Animal Sciences University,
Thoothukkudi-628 008, India
Email: rsanthaanamin@yahoo.co.in

Books in the Series:

- Biology and Culture of Portunid Crabs of the World Seas
- Biology and Ecology of Edible Marine Bivalve Molluscs
- Biology and Ecology of Edible Marine Gastropod Molluscs
- Biology and Ecology of Venomous Marine Snails
- Biology and Ecology of Venomous Stingrays
- Biology and Ecology of Toxic Pufferfish

CONTENTS

LIST OF ABBREVIATIONS

DW	disc width
GFEMN	Global Fisheries Ecosystem Management Network
IUCN	International Union for Conservation of Nature
SMZ	sulfamethoxazole
STL	spine total length
TL	total length
TMAO	trimethylamine oxide
TMP	trimethoprim

PREFACE

A total of 218 species of stingrays have so far been described, including 177 species of marine stingrays with 19 genera and 41 species of freshwater stingrays with seven genera. While marine stingrays have been found widely distributed throughout the coastal *tropical* and *subtropical* seas of the world, freshwater stingrays (family: Potamogtrygonidae) inhabit the brackish waters, lagoons, and freshwater tributaries of some of the major tropical river systems of South America and Africa. The marine stingrays include whiptail stingrays (Dasyatidae); stingrees or round stingrays (Urolophidae); eagle, bull, devil, manta, and cownose rays (*Myliobatidae*); butterfly rays (*Gymnuridae*), sixgill stingrays (*Hexatrygonidae*); and deepwater stingrays (*Plesiobatidae*). Among these marine stingrays, the dasyatid and urolophid stingrays along with potomotrygonid rays have been reported to cause the majority of venomous stings in humans.

Envenomations caused by stingrays are relatively common in fishing communities either from the seas or rivers. In the United States alone, 750–1500 stingray injuries have been reported per year. As more vacationers spend their leisure time exploring coasts and tropical reefs, often in isolated areas without immediate access to advanced health care, there will be greater potential for stingray injuries. A thorough understanding about the diversity of stingrays of marine and freshwater ecosystems and their injuries and envenomations would largely improve the public health community's ability to better manage and to prevent stingray injuries.

Although several books on hazardous marine animals are available, a comprehensive book on the venomous stingrays of the world seas and freshwater systems is overdue. Aspects such as biology and ecology of marine and freshwater stingrays; profiles of world's 220 species of marine and freshwater stingrays and stingray injuries; and their management and treatment are dealt with in this publication. For each species, its common name, global distribution, habitat(s), identifying features, food habits, reproduction, predators, parasites and IUCN's conservation status have been given with suitable illustrations.

It is hoped that this publication would be of great use for the students of fisheries science, marine biology, aquatic biology, and environmental sciences; as a standard reference for libraries of colleges and universities; and as a guide for sea goers and divers.

I am highly indebted to Dr. K. Venkataramanujam, former Dean, Fisheries College and Research Institute, Tamil Nadu Veterinary and Animal Sciences University, Thoothukudi, India, for his valuable suggestions. I am grateful to all my international friends who provided me with certain fish photographs. I also sincerely thank Mrs. Albin Panimalar Ramesh for her help in secretarial assistance and photography.

CHAPTER 1

INTRODUCTION

CONTENTS

ABSTRACT

Stingrays (phylum: Chordata; subphylum: Vertebrata; class: Chondrich-thyes; subclass: Elasmobranchii; order: Myliobatiformes) are members of the "cartilaginous fishes," which are characterized by cartilaginous skel-etons. They have jaws, paired fins and nostrils, scales, and two-chambered hearts. The habitat and distribution of marine and freshwater stingrays along with their use to humans in terms of food, various products, and ecotourism are given in this chapter.

1.1 BIODIVERSITY AND DISTRIBUTION OF WORLD STINGRAYS

A total of 218 species of stingrays have so far been described including 177 species of marine stingrays with 19 genera and 41 species of fresh-water stingrays with 7 genera. While marine stingrays have been found widely distributed throughout the coastal tropical and subtropical seas of the world, freshwater stingrays (family: Potamotrygonidae) inhabit the brackish waters, lagoons, and freshwater tributaries of some of the major tropical river systems of South America and Africa. Most myliobatoid rays are demersal and the eagle rays are pelagic (Michael, 2005). The dasy-atid stingrays, on the other hand, are bottom-feeders in shallow brackish waters and near reefs.

Biodiversity and Distribution of World Stingrays

Family/genera	No. of species	Distribution
Dasyatidae (whiptail stingrays)		
Dasyatis	43[a]	Atlantic, Indian, and Pacific Oceans
Himantura	28[b]	
Makararaja	1[c]	
Neotrygon	5	
Pteroplatytrygon	1	
Pastinachus	5	
Taeniura	3	
Taeniurops	1	
Urogymnus	1	

Family/genera	No. of species	Distribution
Potamotrygonidae (freshwater stingrays)		
Heliotrygon	2[c]	Atlantic and Caribbean watersheds of South America and rivers in West Africa
Paratrygon	1[c]	
Plesiotrygon	2[c]	
Potamotrygon	24[c]	
Gymnuridae (butterfly rays)		
Gymnura	14	Worldwide in tropical and warm temperate (subtropical) seas; Atlantic (Black Sea), Pacific, and Indian oceans
Myliobatidae (eagle rays/manta rays)		
Aetobatus	3	Tropical and western temperate seas worldwide; Atlantic, Pacific, and Indian oceans
Aetomylaeus	4	
Myliobatis	12	
Manta	2	
Mobula	9	
Pteromylaeus	2	
Rhinopteridae (cownose rays)		
Rhinoptera	8	Circumglobal distribution (temperate and tropical continental seas)
Urolophidae (stingarees or round stingrays)		
Trygonoptera	6	Eastern Indian, western Pacific, eastern Pacific (from California to Chile), and western Atlantic ocean
Urolophus	22	
Urobatis	6	
Urotrygon	13	
Hexatrygonidae (sixgill stingrays)		
Hexatrygon	1	Off South Africa
Plesiobatidae (deepwater stingray)		
Plesiobatis	1	South Africa; Mozambique, Australia and Western Indian ocean; west-central Pacific ocean (from Japan to Philippines) and Hawaiian Islands

[a]Four species of *Dasyatis* and [b]seven species of *Himantura* are in freshwater habitats; [c]exclusively freshwater species.

Source: Schneider (1990).

1.2   FOOD USES OF STINGRAYS

Stingrays are of use to humans in terms of food, various products, and ecotourism. Proximate composition of several species of marine stingrays showed that they can be used as a food item like their counterparts, namely, bony fish. Among the marine stingrays, *Dasyatis americana*, *Dasyatis sabina*, *Dasyatis zugei*, and *Dasyatis pastinaca* are suitable for our daily diet. Among freshwater stingrays, *Potamotrygon orbignyi*, *Potamotrygon scobina*, *Potamotrygon motoro*, *Paratrygon aiereba*, and *Plesiotrygon iwamae* are mainly caught for food purposes. Although edible, stingrays are not considered as a high-quality food. They are consumed fresh, dried, and salted. Stingray recipes particularly with dried forms of the wings are most common throughout the world. Normally, the most prized parts of the stingray are the wings, the cheek (the area surrounding the eyes), and the liver. In Singapore and Malaysia, the stingrays are commonly barbecued over charcoal and served with spicy sambal sauce.

1.3   OTHER USES OF STINGRAYS

Among the stingrays, the freshwater species are preferred in the South American ornamental fish trade. While species such as *Plesiotrygon iwamae*, *Paratrygon aiereba*, *Potamotrygon motoro*, *P. orbignyi*, and *P. scobina* are used for ornamental and medicinal purposes, species like *Potamotrygon hystrix* and *P. schroederi* are employed mainly in the ornamental fish trade. The money generated by the freshwater stingray fisheries provides a precious supplementary income for the riverine people. The skin of the stingrays is used as a leather wrap in Japanese swords due to its hard, rough texture. Native American Indians used the spines of stingrays for arrowheads, while groups in the Indo-West Pacific used them as war clubs.

1.4   ECOTOURISM

In the marine sanctuary of the Belize, off the island of Ambergris Caye, divers and snorkelers often gather to watch stingrays. Many Tahitian island resorts regularly offer guests the chance to "feed the stingrays." This consists of taking a boat to the outer lagoon reefs, where habituated stingrays swarm around, pressing right up against tourists seeking food from their hands or that being tossed into the water.

KEYWORDS

- **cartilaginous fishes**
- **marine stingrays**
- **freshwater stingrays**
- **ecotourism**
- **ornamental fish trade**

CHAPTER 2

BIOLOGY AND ECOLOGY OF MARINE STINGRAYS

CONTENTS

ABSTRACT

The ecology, morphology, and internal anatomy of marine stingrays along with their predators, threats, and conservation status are dealt with in this chapter.

2.1 ECOLOGY OF MARINE STINGRAYS

Marine stingrays prefer warm waters. They live in temperate and tropical ocean zones, including open bays and regions near coastlines. They may be divided into categories depending on the ocean depths they occupy. Pelagic stingrays, such as the manta ray, swim actively through open waters. More common, however, are benthic stingrays such as the southern stingray, which swim along the ocean floor and even hide in the sand.

2.2 BIOLOGY OF MARINE STINGRAYS

2.2.1 *CHARACTERISTICS OF MARINE STINGRAY FAMILIES*

2.2.1.1 *FAMILY: DASYATIDAE (WHIPTAIL STINGRAYS)*

The stingrays of this family are chiefly marine and some species are, however, found either in brackish waters or in freshwater systems. The side of the head in these fishes is continuous with the anterior margin of the pectoral fin. These fishes respire by drawing water through a small hole located behind the eye and expelling it through the gill slits on the underside of the disk. The dorsal fin is totally absent or indistinct, when present. The disk is about 1.2 times as broad as long. The eyes and spiracles are located on top of the head. The floor of the mouth is with fleshy papillae. Small teeth are present in many series forming bands along jaws. The nasal curtains are well developed and are continuous across narrow isthmus in front of the mouth and deeply fringed. There is no caudal fin. The tail is long and whiplike. Most species possess at least one long venomous spine on the tail. The dorsal surface is usually gray to dark brown, sometimes with darker or paler markings and the ventral surface is generally whitish. Largest species is about 4 m in length or width. These

fishes are live bearing (ovoviviparous) with fully developed young (ftp://
ftp.fao.org/docrep/fao/009/y4160e/y4160e39.pdf).

2.2.1.2   FAMILY: GYMNURIDAE (BUTTERFLY RAYS)

These fishes are chiefly marine and are rarely seen in estuaries. The body
of these rays is flattened and is surrounded by an extremely broad, rhomboid disk formed by the pectoral fins, which merge in front of the head.
The eyes and spiracles are located on top of the head. Some species have
spiracular tentacles. The snout is obtuse and angular. The nasal curtains
are broadly expanded; continuous across narrow isthmus in front of mouth
and are smooth edged (with rare exceptions). The mouth is slightly arched
and it lacks papillae on its floor. The jaws bear many small teeth in bands.
The dorsal fin and tail spines are present or absent. The pelvic fins are
laterally expanded and are not divided into anterior and posterior lobes.
The caudal fin is always absent. Some species have one or more long,
serrated spines. The tail is short and threadlike with longitudinal folds on
the upper and/or lower surfaces. The skin of the upper side is naked in
most species, but with a variable number of tubercles in large individuals
of other species. The dorsal surface is gray, light green, olive, purple, or
dark brown, sometimes with a reddish cast, often marked with spots or
lines. The ventral surface is white, sometimes with a bronze or rusty cast.
These fishes range in body length from 31 cm to 4 m (ftp://ftp.fao.org/
docrep/fao/009/y4160e/y4160e40.pdf).

2.2.1.3   FAMILY: MYLIOBATIDAE (EAGLE AND MANTA RAYS)

In the rays of this family, the head is elevated above disk. In the eagle
rays, the jaws are powerful with large platelike crushing teeth in several
rows. The eyes and spiracles are laterally placed on head. The tail is
much longer than the disk. Venomous spine(s) is present in some species.
The dorsal fin is small and the pectoral fins are reduced or absent. Some
species are known for their leaping ability high into the air. All the species
are viviparous with 2–6 fully developed young. The plankton-filtering
manta ray species are among the largest fishes of this family but they are
harmless.

2.2.1.4   FAMILY: RHINOPTERIDAE (COWNOSE RAYS)

These ray species are known for their odd-looking heads. Their whiplike tails are armed with one or more stings.

2.2.1.5   FAMILY: UROLOPHIDAE (STINGREES OR ROUND STINGRAYS)

The species of this family possess well-developed caudal fin. The tail is moderately long and outer anterior margins of the pectorals are continuous along the side of the head. Most species have one or more long poisonous spines on tail.

2.2.1.6   FAMILY: HEXATRYGONIDAE (SIXGILL STINGRAY)

The sixgill stingray (*Hexatrygon bickelli*) is the only extant member of this *family*. This flabby, heavy-bodied fish is with a long, thick, fleshy snout. The body coloration is dark violet-blue or brownish above and white below.

2.2.1.7   FAMILY: PLESIOBATIDAE (DEEPWATER STINGRAYS)

The deepwater stingrays are found on soft bottoms at depths between 44 and 680 m. They are large, dark rays with a rounded disk which ends in an angular pointed snout. These rays are viviparous and their reproductive characteristics largely resemble that of *Urolophidae*. They feed on a variety of fishes and invertebrates. They are not of very much interest to fisheries. Despite their venomous defensive sting, they do not pose any threat to humans (Nishida &Nakaya, 1990). Characteristics of some of the important groups of marine stingrays are given below:

Cownose ray: This stingray has a very broad disk and pointed wings. Its snout is indented in the middle to form two lobes, hence the name "cownose." Superficially, it resembles the eagle ray. The eyes are located in front, or anterior to, the beginning of the pectoral fins. Its maximum width and weight may be 2 m and 50 kg, respectively. It may be seen in large schools in sand flats and mudflats stirring up food on the bottom with its wings or pectoral fins.

Southern stingray: This is an inshore species which spends much of its time in shallow areas of sand or mud in search of food. The diet of this ray consists of clams, crabs, shrimps, worms, and small fish. Its disk is rhomboid in shape but it is distinguished largely by its blunt or rounded snout. The tail is rather long and whiplike with a barbed spine near the base.

Atlantic stingray: This has a prominently pointed snout. Its dorsal surface is brown to yellowish-brown and its ventral side is whitish. It is a small species, growing to only 7 kg.

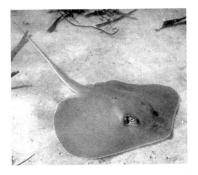

Smooth butterfly ray: This ray has a very broad disk which is much wider than long. The tail is very short with a keel on its top. The tail spine is absent. This ray is unique because it can change its color to match with the sandy bottom of the sea. It can grow to 1.5 m in width and is found in shallow and deep waters. It is most often seen in late spring and summer in warm waters.

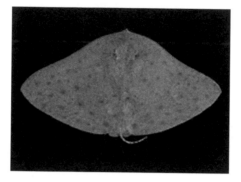

Spotted eagle ray: This has a wide, diamond-shaped disk with whitish, yellow, or green spots on the dorsal side. The ventral side is relatively light in color. One or more sharp barbs are present at the base of a very long, black, whiplike tail. It grows to a width of 2.5 m and a weight of 250 kg. These rays may be seen swimming alone, in pairs, or in schools. They have a shovel-shaped mouth with a wide, single row of teeth that allow them to dig up clams, oysters, and other organisms. Like some other large rays, they have been seen leaping clear out of the water and making loud, croaking sounds.

Atlantic manta: This is the largest ray reaching a width of 7 m, a length of 5 m, and a weight of about 2 t. It has two hornlike projections at the front of the head, which can help push food toward the mouth. These large creatures may be seen basking near the surface of the water. While adult manta rays are known to feed on shrimp, mullet, and plankton, juvenile manta rays feed mainly on anchovies, shrimp, and copepods. Manta rays differ from most other rays in not having a stinging barb.

Devil ray: The devil ray resembles largely the manta ray. But it is considerably smaller, reaching 1.5 m in width.

Morphology: The marine stingrays are related to sharks, skates, and chimeras and are with skeletons made entirely of cartilage. Their body is normally dorsoventrally depressed or "flattened" from top to bottom and the flattened portion of the body is referred to as the "disk." The pectoral fins are large and expanded laterally, becoming thin toward the outer edges to form winglike structures. These fins not only give the ray its unique appearance but also help in locomotion. The pelvic fins are also expanded laterally with a convex lateral margin that is partially overlapped by the pectoral fins. While the dorsal and anal fins are absent, the caudal fin tapers into a filament. The eyes and spiracles are located dorsally. The nictitating membrane is absent and the cornea is attached directly to the skin around the eyes. The gill openings and mouth (with usually protrusible jaws) are on the ventral surface of the body. Unlike bony fish, the stingrays do not have a gill covering. Instead, water flows over their gills through the gill slits. When stingrays are on the ocean floor, they use their spiracles to bring water in for distribution over the gills. These small openings allow the stingrays to breathe while buried in the sand or when in feeding. Some adult stingrays may be no larger than a human palm, while other species, like the short-tailed stingray, may have a body of 2 m in diameter, and an overall length of about 5 m. The smallest stingrays (stingrees or round stingrays) of the family Urolophidae have a body diameter of 30 cm or less. The giant dasyatid stingrays may have a body diameter of more than 2 m and body weight of over 300 kg. The myliobatid rays, especially the giant manta rays, may also have a disk size of up to 7 m with a weight of about 1350 kg. The coloration of stingrays may vary from gray to bright red in color and be plain or patterned (Schneider, 1990). The body of the sting-rays is covered with dermal denticles which are the modified tiny placoid scales. While the benthic marine stingrays tend to ripple their fins to swim above the seafloor, pelagic rays flap their pectoral fins and fly through the water. While the benthic rays usually have a rounded or diamond-shaped body with barbs that are located near the middle of the tail, the epelagic rays have a batlike appearance and have barbs that are located nearer the pelvic/pectoral fins. The teeth of stingrays are also modified placoid scales, and like the dermal denticles, they are less pronounced than that of most of the shark species (Klimley, 2013).

 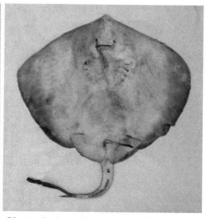

Dorsal view of a marine stingray   **Ventral view of a marine stingray**

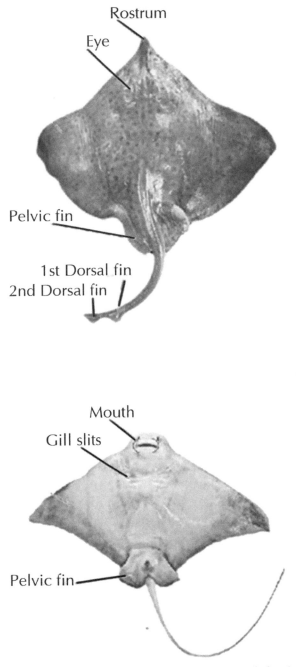

Rostrum

Eye

Pelvic fin

1st Dorsal fin
2nd Dorsal fin

Mouth

Gill slits

Pelvic fin

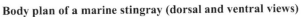

Body plan of a marine stingray (dorsal and ventral views)

2.3 INTERNAL ANATOMY

2.3.1 SKELETON

The internal skeleton of rays (or *endoskeleton*) lacks true bone and is instead made entirely of cartilage. Cartilage is a strong and durable material which is lighter and more flexible than bone, enabling elasmobranchs (which lack a swim bladder) to stay afloat and turn in a tighter radius than other fish. However, parts of the elasmobranch skeleton—such as the skull, the vertebrae, and dermal spines—are often strengthened by the deposition of calcium and salts, a process called *calcification*.

The vertebral centra rays are cylindrical and biconcave in shape and are generally divided into two primary types: precaudal and caudal vertebrae. The number of vertebrae a given individual will contain throughout its entire life is set during embryonic development, a period which also marks the beginning of vertebral calcification. Precaudal vertebrae generally begin forming before caudal vertebrae, and all vertebrae appear to grow throughout the lifespan of individual rays. This enables the use of vertebral centra for ageing. Age is determined in rays by counting vertebral rings that are deposited annually.

2.3.2 LIVER

The liver of rays is a large, soft, and very oily organ which occupies most of the body cavity and can comprise as much as 25% of the body weight. It consists of two large, pointed lobes that are greenish-gray to dark reddish-brown in color. The function of the liver in rays is twofold. First, as in all animals, the liver concentrates the fatty reserves and, therefore, provides for energy storage. Second, the liver acts as a hydrostatic organ by storing lighter-than-water (or low-density) oils. These oils counteract the sinking tendency of the rays by decreasing the density and increasing the buoyancy of the animal on a whole. Without such a large liver, rays would have difficulty staying off the bottom, as they lack the *swim bladder* characteristic of bony fishes.

2.3.3 DIGESTIVE TRACT

The second most noticeable structure in the body cavity of rays is the digestive tract which consists of two contiguous organs: the esophagus and the stomach. The anterior end of the stomach (also known as the *cardiac stomach*) is J shaped and saclike, and tapers into the posterior part of the stomach known as the *pyloric stomach*, which bends anteriorly. The pyloric stomach terminates at a constriction called the *pylorus*, which leads to the short *duodenum* and then to the larger *spiral valve intestine*, which is highly coiled and twisted internally. The function of the spiral valve intestine is to increase the surface area for digestion and absorption of food, while also conserving space in the body cavity for the large liver and for the development of embryos. The spiral valve, in turn, leads to the *rectum* and the *anus*, which opens into the *cloaca*, a cavity where the digestive, urinary, and genital glands open to the outside.

2.3.4 PANCREAS

The pancreas is a gland that helps in digestion by secreting digestive enzymes into the duodenum. It consists of two connected lobes: a ventral lobe, which contains a duct from which pancreatic secretions enter the duodenum, and a dorsal lobe. Both are usually pinkish in color.

2.3.5 SPLEEN

The spleen is a dark brownish organ, triangular or slightly elongate in shape, which lies against the stomach. However, it does not play any role in the digestive process. However, it is part of the lymphatic system, a system which is a major component of the immune system.

2.3.6 RECTAL GLAND

The rectal gland is a small, fingerlike organ that concentrates large quantities of excess salt from the bloodstream for final excretion through the anus. It secretes a colorless solution with about twice the concentration of salt found in the blood plasma into the rectum via a small duct. This organ is very important to rays, whose livers produce large amounts of *urea*, thereby making these marine fishes slightly *hyperosmotic* to seawater).

2.3.7 KIDNEYS

The kidneys of rays are part of the urogenital tract and are involved in the manufacture and transport of urine as well as in the regulation of plasma urea concentrations. They are either semilunar shaped or ribbon-like, dorsoventrally flattened, dark red organs that are highly lobed and lie dorsally on either side of the spinal column outside of the body cavity. A tough membrane, called the *peritoneum*, separates the kidneys from the rest of the body cavity. The kidneys are drained into the cloaca by the ureters.

Food and feeding: The coral reefs are the favorite feeding grounds for the stingrays. The flattened bodies of these rays make them conceal themselves effectively in their environment. They have rows of sensory cells around their mouths called "Ampullae of Lorenzini" that are able to detect weak electric fields generated by prey items. Like sharks, they use these electroreceptors for sensing their prey. Most species of marine stingrays are opportunistic feeders, devouring prey items whenever possible. They feed primarily on small fish, snails, clams, and shrimps, and some other small sea creatures. Many rays are equipped with crusher plates that allow them to crush preys like crabs and shrimps. Some rays, like the manta rays, however, filter-feed on tiny planktonic crustaceans and fish with the help of their transverse gill plates called gill bars. The cephalic (head) lobes of the manta rays help to channel water into the mouth.

Dentition in a typical bluntnose stingray: The upper jaw of this ray protrudes slightly at symphysis, while the lower jaw is indented, leading to a slight overbite. A total of 36–50 rows of teeth are located in the upper jaw. The bottom of the mouth has a cross row of three wide papillae with a lone, small papilla on each side. Each tooth has a quadrangular base. During the mating season, adult male teeth develop wide, triangular cusps for grasping during copulation. Females and juveniles have rounded cusps (Florida Museum of Natural History—Ichthyology).

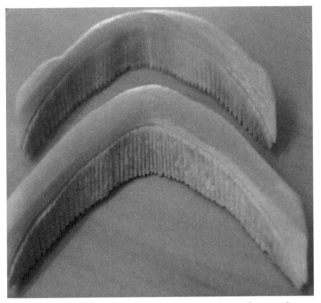

Segments of a tooth plate from the lower jaw of an eagle ray

Behavior: The stingrays, owing to their dorsoventrally flattened body, are well adapted for searching sea floors for food. With muscular wings, they are also hydrodynamically adapted for effortless cruising over long distances along sandy shore lines, often in schools or shoals. They remain motionless; drift along underwater currents; swim backward; and catapult themselves off wave tops, like flying fish. When not feeding or schooling, these rays bury themselves in soft sandy or muddy sea or river bottoms, with their dorsally placed eyes looking for potential prey items or predators. Some freshwater stingrays have been reported to emit electrical currents to stun their prey or predators. When provoked, most stingrays are very capable of defending themselves with their single or multiple tail spines.

2.4   REPRODUCTION (MANTA RAYS)

The reproduction of these fish possesses some similarities with sharks, their close relatives. Manta rays are large fish and just like other species, they develop within eggs; however, the mother does not release them;

it gives birth to live offspring. That is the manta rays are ovoviviparous animals. Fertilization is internal and involves sexual union of two individuals. The male enters one of its copulatory organs called claspers in the female cloaca to transfer its sperm and allow fertilization. It is believed that female manta rays take longer to reach sexual maturity than males. In *Manta alfredi* species, females mature from 8 to 10 years of age and males at 6 years of age, approximately, when the width of the disk is about 2.5–3 m in diameter. *Manta birostris* females also reach maturity between 8 and 10 years or to a later age, while members of the opposite sex mature when their disk's width is 4–4.5 m. The age of maturity varies from region to region. Females usually have 1 or 2 offspring maximum at a time. Manta rays deliver once about every 2–5 years and can have offspring for about 30 years. Of course, half of births may occur during the first 24 or 25 years.

2.4.1 COURTSHIP AND MATING

Mating takes place in warm waters and often around the cleaning stations. When males are in heat, they tend to "casually" wander in the localities in search of a receptive female; the latter releases sex hormones in the water to communicate its willingness to mating. The courtship process may take several days or weeks. Usually, several males (25–30) congregate around a receptive female and compete to mate with her. The female chooses a male and it bites its partner's left pectoral fin to hold her. Then, it positions itself so that bellies of both are bonded and inserts one of its claspers in the female cloaca. The coupling lasts seconds and usually the female stands still. After mating, the male goes away and never returns to take part in parental care.

2.4.2 INCUBATION

After fertilization, the offspring develop in eggs inside the womb. Inside the egg, embryos are fed by the yolk. The female takes care of the eggs for about 1 year until hatching occurs. Then, the young are born alive and independent from the first moment they leave the mother's body. Delivery usually occurs at night and in shallow water.

2.4.3   PATERNAL CARE

Small manta rays can measure more than 1 m in diameter, and since they have few natural predators, they do not need parental care as such. During the first year of life, babies tend to double in size. It is believed that these animals have a long life span which may be at least 40 years (Talwar & Jhingran, 1991; Thorson et al., 1983).

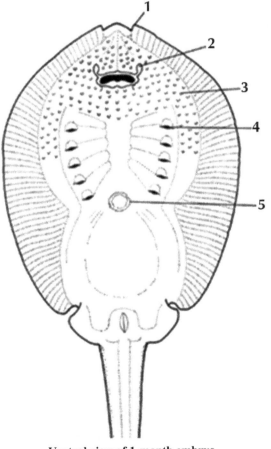

Ventral view of 1-month embryo

(1) Anterior tip of pectoral fin; (2) olfactory pit; (3) papilla; (4) gill opening; (5) point of attachment of yolk.

Predators and other threats: Though the marine stingrays have camouflage counter shading and a sharp barb on their tails, they still encounter a large number of predators. Hammerhead sharks (*Sphyrna* spp.) in particular are voracious consumers of many marine stingray species. The killer whales (*Orcinus orca*) also favor stingray flesh. Other predators include tiger sharks, bull sharks, and large carnivorous fish. Many recreational marine fishermen seek stingrays for their food throughout the world. The flesh of the stingray is often used as a replacement for more valuable fish and crustaceans in seafood salads and premixed seafood entrees.

IUCN conservation status: While most stingrays are relatively widespread and are not currently *threatened*, conservation status for several species (e.g., *Taeniura meyeni*, *Dasyatis colarensis*, *Dasyatis garouaensis*, and *Dasyatis laosensis*) is more problematic and is listed as *vulnerable* or *endangered* by IUCN. Further, the status of several other species of stingrays is poorly known and is listed in the category of *"Data Deficient."*

Conservation strategies for stingrays: The basic conservation objectives for the stingrays are as follows:

1. Securing the biodiversity of stingrays for the future and
2. ensuring the wise, equitable, and sustainable use of stingrays.

Conservation recommendations for stingrays

1. Recognition and protection of aquatic ecosystems of special significance;
2. closed or restricted access to aquatic areas where multiple priority species occur;
3. education of fishers in correct handling and release procedures for priority species;
4. enacting legislation to protecting priority species;
5. assembling the basic biological information for the conservation of genetic diversity;
6. raising awareness, at all levels, of the importance of ecosystems and genetic resource conservation; and
7. training staff to implement the objectives listed above.

KEYWORDS

- **marine stingrays**
- **ecology**
- **morphology**
- **internal anatomy**
- **parental care**
- **threats**
- **conservation**

CHAPTER 3

PROFILE OF MARINE STINGRAYS

CONTENTS

ABSTRACT

The biology, ecology, and parasites of marine stingray species of the different families are given in this chapter.

3.1 WHIPTAIL STINGRAYS (DASYTIDIDAE)

3.1.1 *DASYATIS ACUTIROSTRA* Nishida & Nakaya, 1988

Phylum: Chordata   Subphylum: Vertebrata
Class: Chondrichthyes   Subclass: Elasmobranchii
Order: Myliobtiformes   Family: Dasyatidae

Common name: Sharpnose stingray.

Geographical distribution: Subtropical; northwest Pacific: off the coast of southern Japan and the East China Sea; Gulf of Guayaquil, Ecuador.

Habitat: Marine; demersal.

Distinctive features: Pectoral fin disk of this species is as wide as it is long, with rounded outer margins. A long, triangular snout is present. Eyes are small and are followed by larger spiracles. There is a curtain of skin between the nares, with a fringed, straight posterior margin. Mouth is slightly curved. There are 40–51 upper tooth rows and 39–49 lower tooth rows, arranged in a quincunx pattern. Teeth of adult males have pointed cusps. Pelvic fins are wide and triangular. Tail is whip like and longer than the disk with 1–2 stinging spines on the upper surface. Average spine total length (STL) of males is 39 and 34 mm in females and average total spine serrations in these sexes is 52 and 56, respectively. A subtle dorsal keel and low ventral fin fold are present behind the spine. There is a row of 30 tubercles along the midline of the back and another row of 16 tubercles in front of the spine. Tail behind the spine is covered by small denticles. Coloration is light brown above and white below. The largest known specimen has a disk width (DW) of 72.5 cm (Schwartz, 2007).

Food habits: Virtually nothing is known about the food habits of this species.

Reproduction: It exhibits ovoviparity (aplacental viviparity), with embryos feeding initially on yolk. Additional nourishment for the embryos is received from the mother by indirect absorption of uterine fluid enriched with mucus, fat, or protein through specialized structures (Huveneers, 2006).

Parasites: Not reported.

IUCN conservation status: Near threatened.

3.1.2   *DASYATIS AKAJEI* (Müller & Henle, 1841).

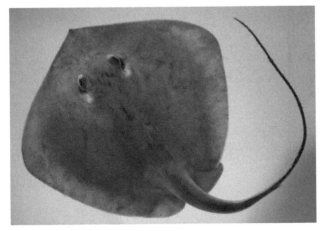

Common name: Red stingray.

Geographical distribution: Endemic to the northwestern Pacific Ocean and is found throughout Japanese coastal waters; Korea, mainland China, and Taiwan; Thailand, Philippines, Fiji, and Tuvalu.

Habitat: Shores and bays; muddy flats, coral reefs, and estuaries.

Distinctive features: This species has a diamond-shaped pectoral fin disk which is wider than long, with nearly straight front margins. A triangular snout is seen. Small eyes are slightly elevated and are followed by spiracles that are almost twice as large. There is a thick flap of skin between the large nares. Teeth are arranged with a quincunx pattern into a pavement-like surface. Females and juveniles have blunt teeth, while adult males have pointed, recurved teeth. There is a row of three papillae across the floor of the mouth. Tail is whip like. A long, serrated spine originates in the first-third of the tail and is followed by a low dorsal keel and a ventral fin fold. Average STLis 42 mm in males and 41 mm in females. The number of total spine serrations in these sexes is 68 and 72, respectively. Adults have a patch of small dermal denticles between and behind the eyes, and a row of thorns along the midline of the back. There are 1–6 tubercles in front of the tail spine, and numerous small denticles behind. Coloration is plain brown above, often with yellow or orange coloring before the eyes, behind the spiracles, around the disk margin, and laterally on the tail in front of the spine. Tail darkens to nearly black toward the tip and on the ventral fin fold. Belly is white with bright orange-red patches. It can grow up to 2 m long and 0.66 m across. Maximum recorded weight is 10.7 kg (Schwartz, 2007).

Food habits: Crustaceans are the most important component of its diet, followed by small bony fishes, annelid worms, and molluscs.

Reproduction: It exhibits ovoviparity (aplacental viviparity), with embryos feeding initially on yolk. Additional nourishment for the embryos is from the mother by indirect absorption of uterine fluid enriched with mucus, fat, or protein through specialized structures. Litter size (number of pups) varies from 1 to 10. Males and females mature sexually at a DW of 35–40 and 50–55 cm, respectively.

Parasites

Monogenea: *Dendromonocotyle akajeii*

Heterocotyle chinensis

Cestoda: *Acanthobothrium dasybati*

Acanthobothrium ijimai

Acanthobothrium grandiceps

Acanthobothrium latum

Acanthobothrium macrocephalum

Acanthobothrium micracantha

Nybelinia aequidentata

Phyllobothrium bifidum

Pterobothrium lintoni

Tetragonocephalum akajeinensis

Nematoda: *Terranova amoyensis*

Copepoda: *Eudactylina dasyati*

Hirudinea: *Rhopalobdella japonica*

IUCN conservation status: Near threatened.

3.1.3   DASYATIS AMERICANA, Hildebrand & Schroeder, 1928

Common name: Southern stingray.

Geographical distribution: Western Atlantic, from New Jersey to Florida, throughout the Gulf of Mexico, Bahamas, and the Greater and Lesser Antilles; south to south-eastern Brazil; Florida and the Bahamas.

Habitat: Sea beds; shallow coastal and estuarine waters and buries itself in sandy bottoms, and occasionally muddy bottoms.

Distinctive features: Its flattened, diamond-shaped body disk has sharp corners, making it more angular. Top of the body varies between olive brown and green and the belly is predominantly white. Pectoral fins are wing like. Tail is slender and it has a long, serrated, and poisonous spine at the base. Eyes are situated on top of the head. Its maximum total length and weight are 152 cm and 97 kg, respectively.

Food habits: Feeds on invertebrates and small fishes.

Reproduction: It exhibits ovoviparity (aplacental viviparity), with embryos feeding initially on yolk. Additional nourishment for the embryos is from the mother by indirect absorption of uterine fluid enriched with mucus, fat, or protein through specialized structures. Gestation period is from 135 to 226 days and litter size varies from 2 to 10.

Parasites

Monogenea: *Dendromonocotyle octodiscus*

Heterocotyle Americana

Thaumatocotyle longicirrus

Thaumatocotyle retorta

Cestoda: *Acanthobothrium americanum*

Anthocephalum alicae

Anthocephalum cairae

Anthocephalum kingae

Lecanicephalum peltatum

Parachristianella monomegacantha

Phyllobothrium centrurum

Polypocephalus medusia

Prochristianella hispida

Pterobothrium kingstoni

Pterobothrium lintoni

Rhinebothrium corymbum

Rhinebothrium lintoni

Rhinebothrium maccallumi

Rhinebothrium magniphallum

Rhinebothrium margaritense

Rhinebothrium spinicephalum

Rhodobothrium pulvinatum

IUCN conservation status: Data deficient.

3.1.4 *DASYATIS BENNETTII* (Müller & Henle, 1841)

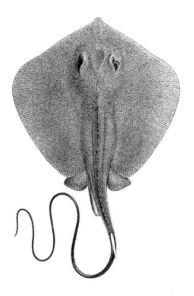

Common name: Bennett's stingray or frill-tailed stingray.

Geographical distribution: Indo-Pacific region (India, through Indo-China, to southern China, Japan, and perhaps the Philippines); north-western Pacific, Vanuatu and New Caledonia.

Habitat: It is a bottom-dweller; coastal waters and freshwaters.

Distinctive features: This species has a diamond-shaped pectoral fin disk which is almost as wide as long, with straight leading margins. A triangular, moderately protruding snout is seen. Trailing margins of the disk are convex. There are 31 upper and 33 lower tooth rows and three or five papillae across the floor of the mouth. Tail is whip like and measures three times the length of the disk. There is a stinging spine on the upper surface of the tail, and a fin fold underneath. Average STL in both males and females is 51 mm. Total spine serrations in these sexes are 95. Juveniles have small dermal denticles in the middle of the back, whereas adults have a row of tubercles along the midline of the back and tiny thorns covering the tail. Coloration of body is yellowish brown above, becoming darker on the tail fold, and light below. This species attains a DW of 50 cm and a total length of 1.3 m (Schwartz, 2007).

Food habits: Preys mainly on fish.

Reproduction: It exhibits ovoviparity (aplacental viviparity), with embryos feeding initially on yolk. Additional nourishment for the embryos is from the mother by indirect absorption of uterine fluid enriched with mucus, fat, or protein through specialized structures.

Parasites: Not reported.

IUCN conservation status: Data deficient.

3.1.5 *DASYATIS BREVICAUDATA* (Hutton, 1875)

Common name: Short-tail stingray or smooth stingray.

Geographical distribution: Temperate waters of the Southern Hemisphere; off southern Africa (from Cape Town in South Africa to the mouth of the Zambezi River in Mozambique); southern Australian coast

Habitat: Bottom living; brackish estuaries, sheltered bays and inlets, sandy flats, rocky reefs, and the outer continental shelf.

Distinctive features: Body of this species is heavily built and smooth. Pectoral fin disk has a rather angular, rhomboid shape and is slightly wider than long. Leading margins of the disk are very gently convex, and converge on a blunt, broadly triangular snout. Eyes are small and are immediately followed by much larger spiracles. Mouth is modestly sized and it has an evenly arched lower jaw, prominent grooves at the corners, and 5–7 papillae (nipple-like structures) on the floor. Teeth are arranged with a quincunx pattern into flattened surfaces. Each tooth is small and blunt, with a roughly diamond-shaped base. There are 45–55 tooth rows in either jaw. Pelvic fins are somewhat large and rounded at the tips. Tail is shorter than the disk and bears one or two serrated stinging spines on the upper surface. Dermal denticles are only found on the tail, with at least one thorn appearing on the tail base. Dorsal coloration is grayish-brown, which darkens toward the tip of the tail and above the eyes. Belly is whitish and is darkening toward the fin margins and beneath the tail. It may reach a size of 2.1 m in DW, 4.3 m in total length, and 350 kg in weight.

Food habits: Feeds primarily on benthic bony fishes and invertebrates, such as molluscs and crustaceans.

Reproduction: It is aplacental viviparous. Once the developing embryos exhaust their yolk supply, they are provided with uterine milk (enriched with proteins, lipids, and mucus) produced by the mother and delivered through specialized extensions of the uterine epithelium called "trophonemata." Female bears a litter size of 6–10. Newborns measure 32–36 cm across.

Predators: Not reported.

Parasites

Cestoda: *Acanthobothrium adlardi*

Dolifusiella ocallaghani

Eutetrarhynchus ocallaghani

Prochristianella mooreae

Pterobothrium lintoni

Trimacracanthus aetobatidis

IUCN conservation status: Least concern.

3.1.6 *DASYATIS BREVIS* (Garman, 1880)

Common name: Whiptail stingray, diamond stingray, Hawaiian stingray.

Geographical distribution: Subtropical; Eastern Pacific: Hawaii and from California, USA to Peru.

Habitat: Marine; demersal; bays, sea grass beds, kelp beds, and near reefs on sand and mud bottoms.

Distinctive features: This species largely resembles *Dasyatis dipteura*. Its characteristic features are the presence of two to five serrations on sides and bending up of spine end. It has a maximum total length of 187 cm and maximum DW 100 cm.

Food habits: Feeds on small fishes, crabs, clams, and other benthic invertebrates.

Reproduction: This species exhibits ovoviparity (aplacental viviparity), with embryos feeding initially on yolk. Additional nourishment for the embryos is from the mother by indirect absorption of uterine fluid enriched with mucus, fat, or protein through specialized structures. Gestation period is between 2 and 2.5 months. Reproduction is annual and litter size ranges from one to four pups. Size at birth ranges from 18 to 23 cm DW and median size at maturity is 58.5 cm DW (males) and 43.4 cm DW (females).

Predators: Not reported.

Parasites

Monogenea: *Listrocephalos kearni*

Cestoda: *Acanthobothrium bullardi*

Acanthobothrium dasi

Acanthobothrium rajivi

Acanthobothrium soberoni

Anthocephalum currani

Pseudochristianella elegantissima

Trematoda: *Probolitrema richiardii*

IUCN conservation status: Data deficient.

3.1.7 *DASYATIS CENTROURA* (Mitchill, 1815)

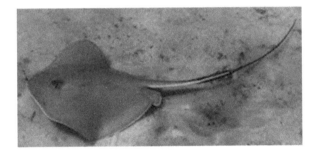

Common name: Roughtail stingray.

Geographical distribution: Subtropical; Eastern Atlantic: southern Bay of Biscay to Angola, including the Mediterranean Sea, Madeira, and Canary Islands; Western Atlantic: Georges Bank to the eastern Gulf of Mexico; southern Brazil to Argentina.

Habitat: Marine; brackish; demersal; sandy and muddy bottoms.

Distinctive features: Disk of this species is sub-quadrangular. Snout is blunt. Other characteristics include large size, spacing of mid-dorsal bucklers, and conspicuous tubercles on the outer parts of disk. Tail is with numerous rows of small spines. Ventral fin fold is long, but quite low. Body coloration is olive-brown above and white below. Lower surface is without dark edgings. Its maximum size and weight are 220 cm (DW) and 300 kg, respectively (Bauchot, 1987).

Food habits: Feeds on bottom-living invertebrates and fishes.

Reproduction: It exhibits ovoviparity (aplacental viviparity), with embryos feeding initially on yolk. Additional nourishment for the embryos is from the mother by indirect absorption of uterine fluid enriched with mucus, fat, or protein through specialized structures. Gestation period is about 4 months with a litter size of 2–4.

Predators: Sharks and other large fishes, particularly *Sphyrna mokarran*, the great hammerhead shark.

Parasites

Monogenea: *Dendromonocotyle centrourae*

Heterocotyle minima

Thaumatocotyle dasybatis

Cestoda: *Acanthobothrium cairae*

Acanthobothrium sp.

Acanthobothrium woodsholei

Anthocephalum centrurum

Dolifusiella tenuispinis

Dolifusiella sp.

Eutetrarhynchus ruficollis

Grillotia erinaceus

Grillotia sp.

Heteronybelinia robusta

Hexacanalis smythii

Kowsalyabothrium indirapriyadarshinii

Lecanicephalum peltatum

Lecanicephalum coangustatum

Oncomegas wageneri

Phyllobothrium foliatum

Prochristianella hispida

Prochristianella sp.

Pterobothrium heteracanthum

Pterobothrium kingstoni

Pterobothrium lintoni

Pterobothrium senegalense

Rhinebothrium maccallumi

Rhodobothrium pulvinatum

Copepoda: *Trebius nunesi*

IUCN conservation status: Least concern.

3.1.8 *DASYATIS CHRYSONOTA* (Smith, 1828)

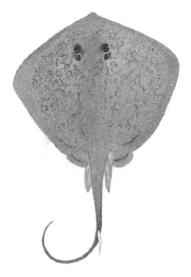

Common name: Blue stingray.

Geographical distribution: Tropical; Southeast Atlantic: central Angola to St. Lucia, Natal, South Africa; Mozambique, Comoros, Reunion, and Madagascar.

Habitat: Marine; demersal; shallow bays, sheltered sandy beaches; deeper waters.

Distinctive features: Disk of this species is rhomboidal and naked except for some denticles near front margin. Snout is angular. Ventral fin fold on tail is falling well short of tail tip. Preorbital snout is longer than the distance between spiracles. Dorsal coloration is golden brown with irregular finger-like blue markings. Belly is white. It grows to a maximum length of 75 cm DW and weight of 25 kg. While males live for 9 years, females live for 14 years (Heemstra & Heemstra, 2004).

Food habits: Feeds on bony fishes and crustaceans such as crabs, shrimps, and mantis shrimps.

Reproduction: It exhibits ovoviparity (aplacental viviparity), with embryos feeding initially on yolk. Additional nourishment for the embryos is from the mother by indirect absorption of uterine fluid enriched with mucus, fat, or protein through specialized structures. Females mature at 7 years of age and 50 cm DW and males at 40.8 cm DW and 5 years. Females live for 14 years and males for 9 years. Females give birth to a litter of 1–5 pups after a gestation period of about 9 months. Size at birth is 17–20 cm DW (Cowley & Compagno, 1993).

Predators: Not reported.

Parasites

Cestoda: *Anthobothrium variabile*

Dollfusiella aculeata

Inermiphyllidium pulvinatum

Onchobothrium pseudouncinatum

IUCN conservation status: Least concern.

3.1.9 *DASYATIS COLARENSIS,* Santos et al., 2004

Image not available.

Common name: Colares stingray.

Geographical distribution: Tropical; southwest Atlantic: Brazil.

Habitat: Marine; pelagic–neritic.

Distinctive features: Disk of this species is rhomboid. Snout is elongated. Lower lip outline is with a dark, well-defined band. Shoulder region to base of tail has a row of small tubercles along the midline of disk. Tubercles are randomly distributed. Posterior margin of the pectoral fins is uniformly rounded. Pelvic fins are triangular shaped. Dorsal caudal keel may be absent or vestigial. It grows to a maximum total length of 207 cm (Santos et al., 2004).

Food habits: Feeds on a wide array of food items.

Reproduction: It exhibits ovoviparity (aplacental viviparity), with embryos feeding initially on yolk. Additional nourishment for the embryos is from the mother by indirect absorption of uterine fluid enriched with mucus, fat, or protein through specialized structures.

Predators: Not reported.

Parasites: Not reported.

IUCN conservation status: Vulnerable.

3.1.10 *DASYATIS DIPTEURA* (Jordan & Gilbert, 1880)

Common name: Diamond stingray.

Geographical distribution: Tropical and warm temperate; eastern Pacific Ocean (southern California to northern Chile, and Galápagos and Hawaiian Islands); Baja California and Gulf of California.

Habitat: Bottom inhabitant of inshore waters; favors sandy or muddy bottoms, often near rocky reefs or kelp forests; intertidal zones.

Distinctive features: Its pectoral fin disk is rhomboid in shape, slightly wider than long, with angular outer corners and subtly convex margins. Snout is blunt-angled and non-projecting. Eyes are fairly large and are immediately followed by spiracles. Mouth is strongly curved, containing

21–37 upper tooth rows and 23–44 lower tooth rows. Teeth are small and blunt and arranged into flattened surfaces. Three or five papillae are found in a row across the floor of the mouth. Whip-like tail measures up to one and half times the length of the disk and bears one long, slender, serrated spine on the upper surface. Behind the spine, there are long dorsal and ventral fin folds. Young rays have completely smooth skin, while adults develop a row of low tubercles along the midline of the back, flanked by two shorter rows on the shoulders. Tail is covered with prickles. Coloration of this species is uniformly olive to brown or gray above, darkening to black on the tail, and off-white below. This species has the lowest growth rate and attains a DW of 1 m. Females generally grow larger than males.

Food habits: Feeds on crustaceans, molluscs, and other invertebrates, as well as small bony fishes.

Reproduction: It is aplacental viviparous and the embryos are initially nourished by yolk, and later by uterine milk (rich in proteins and lipids) produced by the mother. Only the left ovary and uterus are functional in adult females. Males reach sexual maturity at 43–47 cm across and 7 years of age and females at 57–66 cm and 10 years, respectively. Maximum life span has been estimated as 19 years for males and 28 years for females. Newborns measure 18–23 cm across. .

Predators: Not reported.

Parasites

Monogenea: *Listrocephalos kearni*

Cestoda*: Acanthobothrium bullardi*

Acanthobothrium dasi

Acanthobothrium rajivi

Acanthobothrium soberoni

Anthocephalum currani

Parachristianella tiygonis

Pseudochristianello elegantissima

Anaporrhutum euzeti

Probolitrema mexicana

IUCN conservation status: Data deficient.

3.1.11   DASYATIS FLUVIORUM, Ogilby, 1908

Common name: Estuary stingray.

Geographical distribution: Tropical; western Pacific: New Guinea and from the Northern Territory to northern New South Wales in Australia.

Habitat: Marine; brackishwaters; benthopelagic; mangrove swamps and estuaries.

Distinctive features: This species has a diamond-shaped pectoral fin disk which is as wide as long, with gently convex anterior margins and broadly rounded outer corners. Snout is wide and triangular and tapers to a point. Small, widely spaced eyes are immediately followed by the spiracles. Small, bow-shaped mouth is surrounded by deep furrows and it has a row of five papillae across the floor. Teeth are small and arranged into pavement-like surfaces. There are five pairs of gill slits beneath the disk. Pelvic fins are relatively large. Tail measures twice as long as the disk and is broad and flattened at the base. On its upper surface, there are one or two serrated stinging spines. After the spines, tail immediately tapers to become whip like and bears a well-developed keel above and a long, low fin fold beneath. There are wide patches of small dermal denticles with flattened crowns between the eyes and over the middle of the back, along with a midline row of enlarged thorns. Coloration of this species is yellowish to greenish-brown above and lightening toward the disk margins and darkening past the tail spine. Belly is white. It grows to at least 93 cm across and reaches a DW of 1.3 m. Its maximum recorded weight is 6.1 kg.

Food habits: Feeds on crustaceans and polychaete worms.

Reproduction: It exhibits ovoviparity (aplacental viviparity), with embryos feeding initially on yolk. Additional nourishment for the embryos is from the mother by indirect absorption of uterine fluid enriched with mucus, fat, or protein through specialized structures. Newborns measure 11 cm across and 35 cm long. Males mature at 7 years and 41 cm across and females at 13 years and 63 cm across. Maximum life span is 16 years for males and 23 years for females.

Predators: Not reported.

Parasites

Monogenea: *Empruthotrema dasyatidis*

Heterocotyle chinensis

Neoentobdella cribbi

Cestoda: *Callitetrarhynchus gracilis*

Kotorella pronosoma

Parachristianella monomegacantha

Polypocephalus moretonensis

Prochristianella butlerae

Prochristianella mooreae

Shirleyrhynchus aetobatidis

IUCN conservation status: Vulnerable.

3.1.12   DASYATIS GEIJSKESI, Boeseman, 1948

Common name: Sharp snout stingray.

Geographical distribution: Tropical; western Atlantic: northern coast of Venezuela extending southward to Brazil.

Habitat: Marine; brackishwaters; demersal; shallow waters on sandy bottoms.

Distinctive features: The pectoral fin disk of this species is as wide as long, with strongly concave leading margins and rounded corners. Snout is long and projecting. Eyes are minute and are followed by much larger spiracles. Line of the mouth is slightly indented at the center. Upper and lower jaws have 56–68 tooth rows. Teeth are blunt and arranged with a quincunx pattern. A transverse row of five papillae is present on the floor of the mouth. Pelvic fins are distinctive and are twice as long as wide. Whip-like tail measures over twice the length of the disk and bears 1–2 serrated spines on top. Posterior to the spines is a small dorsal keel and a ventral fin fold. A band of small tubercles runs along the dorsal midline from behind the eyes to the base of the tail with larger tubercles in a central row and on each "shoulder." More conical tubercles are scattered over the upper surface of the tail after the spines. Coloration of body is uniformly brown above, and white below darkening toward the disk margin. This species attains a DW of 70 cm.

Food habits: Feeds on small burrowing invertebrates such as worms, crustaceans, and molluscs.

Reproduction: It shows ovoviparity (aplacental viviparity), with embryos feeding initially on yolk. Additional nourishment for the embryos is from the mother by indirect absorption of uterine fluid enriched with mucus, fat, or protein through specialized structures. Females give birth to 1–3 young every year. Younger individuals have generally longer tails (measuring up to three times the DW) than adults.

Predators: Not reported.

Parasites: Not reported.

IUCN conservation status: Near threatened.

3.1.13 *DASYATIS GIGANTEA* (Lindberg, 1930)

Image not available.

Common name: Giant stump tail stingray.

Geographical distribution: Temperate; northwest Pacific: Basargin Cape, Peter the Great Bay and off Askold Island.

Habitat: Marine; demersal.

Distinctive features: This species has a diamond-shaped pectoral fin disk which is much wider than long. Head is short but snout is relatively long and projecting. Eyes are small and widely spaced and are followed by larger spiracles. Pelvic fins are not covered by the disk. Tail is distinctive and is shorter than the disk ending bluntly. There are two stout, saw-toothed spines placed near the base of the tail, and behind them a ventral fin fold. Skin is largely smooth, except for numerous small dermal denticles over the posterior portion of the disk, pelvic fins, and base of the tail. This species grows to a maximum total length of 2.3 m and DW of 1.8 m.

Food habits: It feeds on a wide array of food items.

Reproduction: This species shows ovoviparity (aplacental viviparity), with embryos feeding initially on yolk. Additional nourishment for the embryos is from the mother by indirect absorption of uterine fluid enriched with mucus, fat, or protein through specialized structures.

Predators: Not reported.

Parasites: Not reported.

IUCN conservation status: Data deficient.

3.1.14 *DASYATIS GUTTATA* (Bloch & Schneider, 1801)

Common name: Long nose stingray.

Geographical distribution: Tropical; western Atlantic: southern Gulf of Mexico and West Indies to Santos, Brazil.

Habitat: Marine; demersal; shallow waters.

Distinctive features: This species has a diamond-shaped pectoral fin disk which is slightly wider than long. Anterior margins are concave converges into an obtuse, moderately projecting snout. Mouth is curved with a median projection in the upper jaw which fits into an indentation in the lower jaw. A row of three papillae is found across the floor of the mouth. There are 34–46 tooth rows in the upper jaw. Teeth have tetragonal bases and blunt crowns in females and sharp, pointed cusps in mature males. Pelvic fins are rounded. Slender, whip-like tail is much longer than the disk and usually bears a single serrated stinging spine near the base (some individuals have no spine or more than one). A row of small, blunt thorns or tubercles are present along the midline of the back, from between the eyes to the base of the tail spine. Larger sized rays possess a mid-dorsal band of heart-shaped, flattened denticles. Coloration is olive, brown, or gray above, sometimes with darker spots, and yellowish to white below. Keel and fin folds on the tail are black. This species reaches a maximum known DW of 2 m and females grow larger than males (Stehmann et al., 1978).

Food habits: Feeds mainly on benthic invertebrates and small bony fishes.

Reproduction: This species shows ovoviparity (aplacental viviparity), with embryos feeding initially on yolk. Additional nourishment for the embryos is from the mother by indirect absorption of uterine fluid enriched with mucus, fat, or protein through specialized structures. Females have a single functional uterus, on the left, and bear two litters of 1–2 pups per year. Gestation period is 5–6 months.

Predators: Not reported.

Parasites

Monogenea: *Heterocotyle sulamericana*

Monocotyle guttatae

Cestoda: *Acanthobothrium tasajerasi*

Acanthobothrium urotrygoni

Acanthobothroides thorsoni

Rhinebothrium margaritense

Rhinebothrium magniphallum

Rhodobothrium pulvinatum

IUCN conservation status: Data deficient.

3.1.15   *DASYATIS HASTATA* (DeKay, 1842)

Common name: Roughtail stingray.

Geographical distribution: Tropical to warm temperate waters; north Atlantic; eastern Atlantic (from the southern Bay of Biscay to Angola and in the Mediterranean Sea); western Atlantic (from Georges Bank in Massachusetts to southern Florida and in the northeastern Gulf of Mexico and the Bahamas); off Uruguay and southern Brazil.

Habitat: Marine; brackishwater; demersal; lagoons, shallow bays and estuaries; muddy and sandy substrates.

Distinctive features: This species has a trapezoid-shaped disk with an anterior margin that is straight to slightly concave. Posterior margin of the disk is slightly convex and outer corners are abruptly rounded. Snout is moderately long and angular with an obtuse tip. It lacks a dorsal fin fold, and the ventral fin fold on its tail is long, narrow, and not easily seen. The most distinguishable feature of this species is its tail which contains numerous rows of small thorns and is long, slender, and whip like. Tail is about two and a half times the length of the body. Tubercles are seen on the outer parts of its disk and mid-dorsal bucklers, which are large and widely spaced. Coloration is dark brown or olive color dorsally and white or nearly white ventrally. Tail is black from the spine rearward. The maximum DW and weight of this species are 221 cm and 300 kg, respectively.

Food habits: Feeds on bony fishes, such as scup and sand lance, and of bottom-dwelling invertebrates, such as polychaetes, cephalopods, and crustaceans.

Reproduction: It shows ovoviparity (aplacental viviparity), with embryos feeding initially on yolk. Additional nourishment for the embryos is from the mother by indirect absorption of uterine fluid enriched with mucus, fat, or protein through specialized structures.

Predators: Not reported.

Parasites: Not reported.

IUCN conservation status: Least concern.

3.1.16   *DASYATIS HYPOSTIGMA,* Santos & Carvalho, 2004

Common name: Groovebelly stingray.

Geographical distribution: Tropical; southwest Atlantic: Brazil and Argentina.

Habitat: Marine; brackishwaters and estuaries; benthopelagic; favors sandy or muddy inshore waters.

Distinctive features: It has a diamond-shaped pectoral fin disk which is slightly wider than long, with nearly straight front margins and a barely projecting snout tip. Eyes are large and protruding and are immediately followed by wide, angular spiracles. Mouth is small and lower jaw is strongly bow shaped. There are 37–46 upper and 43–50 lower tooth rows, with a quincunx arrangement. Teeth are blunt in juveniles and females, while in adult males, central teeth are slender and sharp. Adults have three, five, or seven papillae in a row across the floor of the mouth. There is a distinctive "W"-shaped furrow on the underside behind the fifth pair of gill slits. Pelvic fins are pointed with slightly curving trailing margins. Whip-like tail measures up to around one and a half times the disk length, with one (occasionally two) serrated stinging spine positioned on top past the first third of the tail. There are fin folds running along the dorsal and ventral sides of the tail behind the spine. Tiny denticles are found scattered atop the disk around the tail base. Dorsal coloration is yellowish to greenish brown above, becoming more reddish toward the disk margins. Belly is white with dark fin margins, while the tail fin folds are black. This species attains a maximum DW of 65 cm.

Food habits: Feeds on a wide array of food items.

Reproduction: This species is aplacental viviparous. Embryos hatch inside the mother's uterus and are sustained by yolk, later supplemented by uterine milk delivered by the mother into the embryos' spiracles via trophonemata (villi-like structures). Females have a single functional uterus (on the left). It gives birth to two pups only.

Predators: Not reported.

Parasites: Not reported.

IUCN conservation status: Data deficient.

3.1.17 *DASYATIS IZUENSIS,* Nishida & Nakaya, 1988

Common name: Izu stingray.

Geographical distribution: Subtropical; northwest Pacific: Izu Peninsula, Pacific coast of Japan.

Habitat: Marine; demersal; shallow coastal waters.

Distinctive features: Pectoral fin disk of this species is diamond shaped and slightly wider than long, with gently convex forward margins and a blunt snout. Eyes are medium sized and are followed by large spiracles. There are 35–41 tooth rows in the upper jaw and 37–39 tooth rows in the lower jaw. Teeth have a pavement-like quincunx arrangement. They are blunt in females and pointed in mature males. A transverse row of five papillae (nipple-like structures) is found on the floor of the mouth. Pelvic fins are broad and triangular. Tail is whip like, about as long as the disk, and bears one (sometimes two) stinging spines on top. Average STL is 68 mm in males and 79 mm in females and average total number of serrations is 112 in males and 130 in females with 10–33 serrations occurring on the sides of the spine base. Spine has a short groove and typically measures 6.8 cm long in males and 7.9 cm in females with 112 and 130 serrations, respectively. Posterior to the spine, there is a short and slight dorsal keel and a ventral fin fold. Skin is completely smooth excepting 2–6 small tubercles located before the spine in the largest individuals. Coloration is golden brown above, darkening between the eyes and on the last two-thirds of the tail, and belly is white. Ventral tail fold is white. It grows to a maximum size of 42 cm DW (Schwartz, 2007).

Food habits: Feeds on a wide array of food items.

Reproduction: This species shows ovoviparity (aplacental viviparity), with embryos feeding initially on yolk. Additional nourishment for the embryos is from the mother by indirect absorption of uterine fluid enriched with mucus, fat, or protein through specialized structures. It attains sexual maturity at a DW of around 37 cm.

Predators: Not reported.

Parasites: Not reported.

IUCN conservation status: Near threatened.

3.1.18 DASYATIS LAEVIGATA, Chu, 1960

Common name: Yantai stingray.

Geographical distribution: Subtropical; northwest Pacific: China and Taiwan.

Habitat: Marine; demersal; inshore waters and estuaries.

Distinctive features: Pectoral fin disk of this species is diamond shaped and 1.2–1.3 times wider than long, with nearly straight anterior margins converging to a blunt-angled snout, and rounded outer margins. Large, protruding eyes are immediately followed by a pair of equal-sized or slightly smaller, elliptical spiracles. Mouth is small and bow shaped, with three papillae across the floor. Teeth of adult males have a sharp, recurved cusp while those of females are blunt. Five pairs of short gill slits are seen. Pelvic fins are almost rectangular in shape. In females, pelvic fins are wider and more strongly curved posteriorly than those of males. Tail measures 1.4–1.8 times the disk length and is laterally compressed besides becoming whip like toward the tip. Stinging tail spine measures about 42 mm long and bears on average 60 serrations in both males and females. After the spine, there is a short, narrow dorsal fin fold and a wider ventral fin fold running along two-fifths of the tail. Skin is completely devoid of dermal denticles even in adults. Disk is yellowish gray-brown above, with irregular darker blotches and yellow marks adjacent to the eyes and spiracles. Belly is white with irregular darker spots and a yellowish gray margin. Tail is dark brown with a yellow lateral stripe and black fin folds. It is a small species and male and female have a DW of 20 and 30 cm, respectively (Schwartz, 2007).

Food habits: Feeds on a wide array of items.

Reproduction: It shows ovoviparity (aplacental viviparity), with embryos feeding initially on yolk. Additional nourishment for the embryos is from the mother by indirect absorption of uterine fluid enriched with mucus, fat, or protein through specialized structures. Females give birth to only 1–2 pups at a time.

Predators: Not reported.

Parasites: Not reported.

IUCN conservation status: Near threatened.

3.1.19 *DASYATIS LATA* (Garman, 1880)

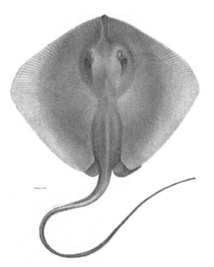

Common name: Brown stingray.

Geographical distribution: Subtropical; circumglobal: Pacific.

Habitat: Marine; demersal; sand or mud bottoms, sometimes near coral reefs.

Distinctive features: It has a diamond-shaped pectoral fin disk which is with straight leading margins that converge at an obtuse angle, and curved trailing margins. Tip of the snout is rounded and protrudes past the disk. Mouth is arched and contains 5–6 papillae on the floor. Pelvic fins are short and rounded. Whip-like tail is twice the length of the disk or more and bears a serrated stinging spine on the upper surface near the tail base. Average SPL of males is 83 and 122 mm in females with 1–13 serrations occurring on the sides of the spine base. There is a long, narrow fin fold behind the tail, which finally becomes a keel that runs all the way to the tail tip. Larger individuals have three large, elongated tubercles in the middle of the back. Tail is roughened by small dermal denticles, along with an irregular row of conical tubercles on each side and several large, flattened tubercles in front of the spine. Coloration of this species is plain olive to brown above and white below. It can reach 1.5 m DW and 56 kg in weight (Schwartz, 2007).

Food habits: Feeds on bottom-dwelling crustaceans, polychaete worms, and small bony fishes.

Reproduction: It shows ovoviparity (aplacental viviparity), with embryos feeding initially on yolk. Additional nourishment for the embryos is from the mother by indirect absorption of uterine fluid enriched with mucus, fat, or protein through specialized structures.

Predators: Scalloped hammerheads (*Sphyrna lewini*) are the dominant predators.

Parasites

Cestoda: *Acanthobothrium chengi*

Acanthobothrium hawaiiensis

Dollfusiella sp.

Parachristianella monomegacantha

Parachristianella sp.

Prochristianella micracantha

Pterobothrium hawaiiense

IUCN conservation status: Least concern.

3.1.20 *DASYATIS LONGA* (Garman, 1880)

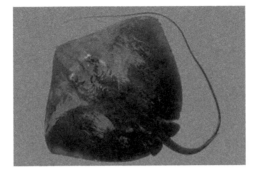

Common name: Longtail stingray.

Geographical distribution: Tropical; eastern Pacific: southern Gulf of California to Colombia and the Galapagos Islands.

Habitat: Marine; rocky and coral reefs; soft and sandy bottoms.

Distinctive features: This species has a diamond-shaped pectoral fin disk which is about one-sixth wider than long, with the outer corners broadly rounded. Front margins are nearly straight, meeting the tip of the snout at a blunt angle. There is a row of five papillae across the floor of the mouth. Pelvic fins are rounded. Whip-like tail bears a stinging spine and measures more than twice as long as the disk. Behind the spine, the tail becomes laterally compressed with a low keel above and a short, narrow fin fold below. There is a row of pointed tubercles running along the midline of the back from between the "shoulders" to the base of the tail. Two much shorter rows of smaller tubercles are seen alongside the central row behind the shoulders. Numerous small dermal denticles are also seen between the eyes and on the tail behind the spine. Dorsal coloration varies from plain reddish-brown to dark gray and belly is light. This species reaches a maximum known DW of 1.58 m, total length of 2.57 m, and weight of 46.4 kg.

Food habits: Predator of bottom-dwelling bony fishes and invertebrates, in particular stomatopods, decapods, and molluscs.

Reproduction: It shows ovoviparity (aplacental viviparity), with embryos feeding initially on yolk. Additional nourishment for the embryos is from the mother by indirect absorption of uterine fluid enriched with mucus, fat, or protein through specialized structures. Gestation period is 10–11 months. A litter contains 1–5 young, each measuring about 40 cm across. Males mature sexually at 0.8 m across and females at 1.1 m across.

Predators: Not reported.

Parasites

Monogenea: *Listrocephalos whittingtoni*
Cestoda: *Acanthobothrium campbelli*
Acanthobothrium costarricense
Acanthobothrium cimari
Acanthobothrium cleofanus
Acanthobothrium obuncus
Acanthobothrium puntarenasense
Acanthobothrium vargasi
Anthocephalum lukei
Anthocephalum michaeli
Pseudochristianella elegantissima
Pseudochristianella nudiskula
Pterobothrioides carvajali
Trematoda: Probolitrema richiardii
IUCN conservation status: Data deficient.

3.1.21 DASYATIS MARGARITA (Günther, 1870)

Common name: Daisy stingray.

Geographical distribution: Tropical; eastern Atlantic: Mauritania to Angola.

Habitat: Marine; brackishwater; demersal; common in marine and estuarine habitats.

Distinctive features: Pectoral fin disk of this species is moderately thin and rounded, measuring about as wide as long. Leading margins of the disk are concave and converge at the pointed, slightly projecting tip of the snout. Eyes are medium-sized and followed by somewhat larger spiracles. There are five papillae in a transverse row across the floor of the mouth. Tooth rows number 24–32 in the upper jaw and 28–36 rows in the lower jaw and are arranged with a quincunx pattern into pavement-like surfaces. Pelvic fins are short, with the tips projecting just past the disk margin. Tail is longer than the disk and usually bears a single long, thin stinging spine on the upper surface. It is broad and flattened at the base, becoming slender and whip like. After the spine, there is a low dorsal keel and a well-developed ventral fin fold. There is a massive, circular pearl spine at the center of the disk. Older rays over 20 cm DW gain a wide band of small, flattened, circular dermal denticles covering the median third of the back from between the eyes to the base of the tail, as well as small prickles covering the tail behind the sting. This species is plain grayish-brown above and whitish below. It reaches a maximum known DW of 1 m and weight of 17 kg.

Food habits: Feeds on shrimps, crabs, bivalves, and annelids.

Reproduction: It shows ovoviparity (aplacental viviparity), with embryos feeding initially on yolk. Additional nourishment for the embryos is from the mother by indirect absorption of uterine fluid enriched with mucus, fat, or protein through specialized structures. Females bear litters of 1–4 pups, with coastal lagoons and estuaries serving as breeding grounds (Capapé & Desoutter, 1990).

Predators: Great hammerheads (*Sphyrna mokarran*) and other larger fish including sharks.

Parasites

Cestoda: *Otobothrium cysticum*

IUCN conservation status: Endangered.

3.1.22   *DASYATIS MARGARITELLA,* Compagno & Roberts, 1984

Common name: Pearl stingray.

Geographical distribution: Western coast of Africa, from Cape Blanc in Mauritania to Angola.

Habitat: Bottom dwelling species; shallow, coastal marine and brackish-waters; lagoons and estuaries.

Distinctive features: This species has a moderately thin, oval-shaped pectoral fin disk which is about as long as wide. Narrow snout tapers to a point that protrudes slightly from the disk. Eyes are immediately followed by the spiracles, which are of approximately equal size. There are five papillae in a row across the floor of the mouth. There are 24–41 upper tooth rows and 34–50 lower tooth rows. Blunt, ridged teeth are arranged into pavements with a quincunx pattern. Pelvic fins are short and trian-gular with their tips projecting slightly beyond the disk margin. Tail is broad and flattened at the base and is becoming thin and whip like after the (usually) single, slender stinging spine on the upper surface. After the spine, there is a low dorsal keel and a well-developed ventral fin fold. There is a medium-sized oval pearl spine in the middle of the back. Individuals of more than 13–14 cm across also gain a band of small, heart-shaped, or flat-tened circular dermal denticles covering the median third of the disk, from between the eyes to the base of the tail. Tail is covered by small prickles behind the spine. Coloration is plain grayish-brown above and completely white below. This species grows to 30 cm across and a weight of 1 kg.

Food habits: Feeds on a wide array of food items.

Reproduction: Like other stingrays, it is aplacental viviparous with females bearing litters of 1–3 pups. Sexual maturation occurs at a DW of around 20 cm.

Predators: Not reported.

Parasites: Not reported.

IUCN conservation status: Data deficient.

3.1.23   *DASYATIS MARIANAE,* Gomes et al., 2000

Common name: Brazilian large-eyed stingray.

Geographical distribution: Northeastern Brazil from Parcel Manoel Luís off Maranhão state to southern Bahia.

Habitat: Coral or sandstone reefs on the continental shelf; Younger rays inhabit nearshore sandy flats and estuaries in addition to reefs.

Distinctive features: This species has a diamond-shaped pectoral fin disk which is approximately as long as wide, with rounded outer corners and slightly concave front margins. Tip of the snout is not strongly projecting. Eyes are large, about equal in width to the spiracles and the distance between the eyes. Mouth is small, with a transverse row of three papillae on the floor. There are 35–45 tooth rows in the upper jaw and 38–48 tooth rows in the lower jaw. The teeth of females are blunt, while those of older males are pointed. Tail measures no longer than 1.5 times the DW and tapers to a filament toward the tip. There is a stinging spine on top of the tail, followed by dorsal and ventral fin folds. Disk is largely smooth, except for a row of 2–18 small, thorn-like dermal denticles along the dorsal midline in adults and a single thorn on each shoulder in males. Coloration of this species is distinctive. Back is golden brown with the disk and pelvic fins are edged by a thin blue line and then a dark brown band. There are dark brown blotches around the eyes, between the spiracles, and in two pairs behind the spiracles and further back on the disk. Adult males have blue coloring on top and at the tip of the claspers. Belly is white, darkening at the disk margin. Tail is brown above and white below, darkening to purple at the tip. Largest specimen measures 40 cm DW.

Food habits: It feeds on a wide array of food items.

Reproduction: Like other stingrays, this species is aplacental viviparous. Females have a single functional uterus (on the left) and carry a single embryo at a time. Embryo is provided by yolk, and later uterine milk is secreted by the mother. Gestation period is 5–6 months, and females are able to bear two litters per year in June and in November and December. Newborn rays measure 13–14 cm across. Females mature later and attain a larger ultimate size than males.

Predator: Cobia (*Rachycentron canadum*).

Parasites: Not reported.

IUCN conservation status: Data deficient.

3.1.24 *DASYATIS MARMORATA* (Steindachner, 1892)

Common name: Marbled stingray.

Geographical distribution: Tropical; eastern Atlantic and Mediterranean Sea: Morocco, Mauritania to Congo; Natal, South Africa.

Habitat: Marine; demersal; close inshore on sandy beaches and in shallow bays; sometimes near rocky reefs; deeper offshore areas.

Distinctive features: Conspicuous bright blue blotches and branching lines are seen on a golden brown disk of this species. Snout and disk are angular. Tail is less than twice body length with a short upper caudal fin fold and a longer lower one falling far in front of tail tip. Disk is without thorns. There is usually one sting. Belly is white. Tail is darker and without bands. It grows to a maximum total length of 60 cm (Cowley & Compagno, 1993).

Food habits: Feeds on crabs, mantis shrimps, amphipods, worms, and fishes.

Reproduction: It shows ovoviparity (aplacental viviparity), with embryos feeding initially on yolk. Additional nourishment for the embryos is from the mother by indirect absorption of uterine fluid enriched with mucus, fat, or protein through specialized structures.

Predators: Not reported.

Parasites

Monogenea: *Dendromonocotyle octodiscus*

Cestoda: *Dolifusiella aculeata*

IUCN conservation status: Data deficient.

3.1.25 *DASYATIS MATSUBARAI,* Miyosi, 1939

Common name: Pitted stingray.

Geographical distribution: Subtropical; northwest Pacific: Japan and South Korea.

Habitat: Marine; demersal; coastal waters.

Distinctive features: This species has a diamond-shaped pectoral fin disk which is wider than long. Floor of the mouth contains from nil papillae to 12. There are 34–44 upper tooth rows and 33–46 lower tooth rows. There is a distinctive "W"-shaped furrow on the underside of the disk, at the center behind the fifth pair of gill slits. Tail is whip like and is bearing 1–3 stinging spines on the upper surface. Tail spine averages 6.5 cm long with 90 serrations in males and 7.7 cm long with 87 serrations in females. There is a low dorsal keel after the spine and a ventral fin fold measuring less than half as long as the DW. Older individuals have a row of 2–10 tubercles on the snout tip, 3–5 tubercles on the back, and 1–8 tubercles before the spine. Tail is covered by dermal denticles toward the tip. Coloration of this species is dark gray above, darkening on the tail fold, and white below with gray irregular spots and fin margins. Upper surface of the disk bears many small pores that are ringed in white. This species attains a DW of 1.2 m (Schwartz, 2007).

Food habits: Fish, bivalves, squid, and crustaceans are the food items of this species.

Reproduction: This species shows ovoviparity (aplacental viviparity), with embryos feeding initially on yolk. Additional nourishment for the embryos is from the mother by indirect absorption of uterine fluid enriched with mucus, fat, or protein through specialized structures.

Predators: Not reported.

Parasites: Praniza larvae of gnathiid isopods which attach to the gills of this species.

IUCN conservation status: Data deficient.

3.1.26   *DASYATIS MICROPS* (Annandale, 1908)

Common name: Smalleye stingray.

Geographical distribution: Indo-West Pacific: India and northern; estuary of the River Ganges; Mozambique.

Habitat: Marine; brackishwaters; demersal; deep waters; coastal waters and river mouths.

Distinctive features: Pectoral fin disk of this species is more than 1.4 times wider than long, with the outer corners forming obtuse angles. Anterior margins of the disk are sinuous and converge on a rounded snout with a slightly projecting tip. Eyes are small and immediately followed by a pair of much larger spiracles. Mouth is wide with five papillae across the floor. Tail is broad and flattened from its base to the stinging spine. Tail spine measures 9.1 cm long and bore 75 serrations in females. About 25 spine serrations occur on the sides of the spine base. After the spine, tail abruptly becomes thin and cylindrical, with a low, thick fin fold running underneath. Upper surface of the disk is covered by many large dermal denticles with star-shaped bases, concentrated on the snout. This species is brown to reddish brown above, darkening toward the tip of the tail, and white below, becoming slightly dusky at the fin margins and on the tail. Dorsal surface bears characteristic large white spots beside the eyes, around the disk center, and in a row on either side about two-thirds of the way to the pectoral fin tips. There is also a row of small white spots on either side of the tail base. It can reach a DW of 2.2 m and a total length of 3.2 m (Schwartz, 2007).

Food habits: It feeds on an array of food items.

Reproduction: It shows ovoviparity (aplacental viviparity), with embryos feeding initially on yolk. Additional nourishment for the embryos is from the mother by indirect absorption of uterine fluid enriched with mucus, fat, or protein through specialized structures. Newborn rays measure 31–33 cm across.

Predators: Not reported.

Parasites: Cestoda: *Oncomegoides celatus.*

IUCN conservation status: Data deficient.

3.1.27   *DASYATIS MULTISPINOSA* (Tokarev, 1959) = *Dasyatis matsubarai*

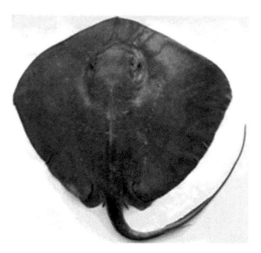

Common name: Multispine giant stingray.

Geographical distribution: Northwest Pacific: Sea of Japan.

Habitat: Marine; bathydemersal; deep waters; semi-pelagic and benthic.

Distinctive features: Pectoral fins of this species are fused with the head, forming a diamond flat disk whose width is greater than the length. Front edge is almost straight, and snout forms an obtuse angle. Behind the eyes spiracles are seen. On the ventral surface of the disk, there are five gill slits, mouth, and nostrils. Mouth is curved in an arc. Teeth are arranged in a staggered manner to form a flat surface. In the mouth, there are 34–44 upper and 33–46 lower tooth rows. There are two or three spines. Tail is studded with scales. Coloration of the dorsal surface of the disk is dark gray. Tail folds of skin are darker than the main background. Ventral surface of the disk is with irregular white spots. Dorsal surface of the disk is covered with numerous pores with white edging. Maximum width of the disk recorded is 1.2 m.

Food habits: It feeds on an array of food items.

Reproduction: It shows ovoviparity (aplacental viviparity), with embryos feeding initially on yolk. Additional nourishment for the embryos is from the mother by indirect absorption of uterine fluid enriched with mucus, fat, or protein through specialized structures.

Predators: Not reported.

Parasites: Parasitic isopods of family gnathiidae have been reported from the gills of this species.

IUCN conservation status: Data deficient.

3.1.28   *DASYATIS NAVARRAE* (Steindachner, 1892)

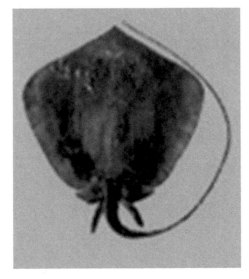

Common name: Blackish stingray.

Geographical distribution: Subtropical; northwest Pacific: Taiwan and Shanghai, China.

Habitat: Marine; demersal; coastal waters and estuaries.

Distinctive features: This species has a diamond-shaped pectoral fin disk which is almost as long as wide, with slightly sinuous leading margins, narrowly rounded outer corners, and almost straight trailing margins. Triangular, projecting snout comprises about one-fourth the disk length and bears 2–3 rows of enlarged pores along the midline. Eyes are small and are closely followed by a pair of larger spiracles. Mouth is bow-shaped, with three papillae across the floor. There are 40 upper tooth rows and 37 lower tooth rows. Teeth of adult males are pointed while those of females are blunt. Tail is whip like and bears a stinging spine on the dorsal surface, as well as both upper and lower fin folds. Ventral fold measures half the disk length. There are 4–6 rows of small tubercles between the eyes, a narrow strip of tubercles running down the center of back to the base of the tail, and three enlarged tubercles in front of the tail spine. Coloration of body is plain dark brown above and whitish below. This species attains a DW of 38 cm and a total length of 60–94 cm (Wang et al., 2009).

Food habits: Very little is known of the habitat and biology of this species.

Reproduction: This species shows ovoviparity (aplacental viviparity), with embryos feeding initially on yolk. Additional nourishment for the embryos is from the mother by indirect absorption of uterine fluid enriched with mucus, fat, or protein through specialized structures.

Predators: Not reported.

Parasites

Monogenea: *Heterocotyle chinensis*

Copepoda: *Caligus dasyaticus*

IUCN conservation status: Data deficient.

3.1.29   *DASYATIS PARVONIGRA,* Last & White, 2008

Common name: Dwarf black stingray.

Geographical distribution: Tropical; Eastern Indian Ocean: Australia; western Pacific (Indonesia, off West Papua and the Philippines).

Habitat: Marine; brackish; benthopelagic; continental shelves.

Distinctive features: It has a diamond-shaped pectoral fin disk which is slightly wider than long, with slight angular outer corners and almost straight anterior corners. Snout is a broadly triangular, with a pointed tip. Eyes are large and elevated and are immediately followed by larger spiracles. Medium-sized mouth forms a strong arch and is with four papillae (nipple-like structures) across the floor and an additional, tiny papilla near the corner of each jaw. Teeth are small and number approximately 43 rows in either jaw. Five pairs of gill slits are slightly S shaped. Pelvic fins are small with nearly straight margins. Males have somewhat flattened claspers. Tail measures around 1.5 times as long as the disk and is broad and flattened at the base. One or two serrated stinging spines are dorsally placed about a third of a DW back from the tail base. After the sting, the tail becomes thin and whip like, bearing a long, low ventral fin fold and a much shorter dorsal ridge. There is a short row of closely spaced, spear-shaped thorns along the midline of the back, starting after the head. One to two small, seed-shaped thorns are also present on each shoulder.

Disk of this species is almost uniformly dark grayish-brown above and is becoming lighter toward the disk margins, on the thorns, and past the sting. Ventral surface of disk and tail is almost uniformly white. It reaches 51 cm across and 1.1 m long (Last & White, http://bionames.org/bionames-archive/issn/1833-2331/22/275.pdf).

Food habits: Not reported.

Reproduction: It is an aplacental viviparous species with the developing embryos sustained on maternally produced uterine milk as in other stingrays. Males attain sexual maturity at around 35 cm across.

Predators: Not reported.

Parasites: Not reported.

IUCN conservation status: Data deficient.

3.1.30 *DASYATIS PASTINACA* (Linnaeus, 1758)

Common name: Common stingray.

Geographical distribution: Subtropical; northeastern Atlantic Ocean; Mediterranean Sea and the African coast.

Habitat: Marine; brackish; demersal; coastal lagoons, shallow bays and estuaries; rocky reefs.

Distinctive features: Pectoral fin disk of this species is diamond shaped and is slightly wider than it is long, with narrowly rounded outer corners. Leading margins of the disk are almost straight and converge on a pointed, slightly protruding snout. Trailing margins of the disk are convex. Eyes are smaller than the spiracles which are placed closely behind. There are 28–38 upper tooth rows and 28–43 lower tooth rows. Teeth are small and blunt and arranged into flattened surfaces. There are five papillae across the floor of the mouth. Tail is slender and whip like, measuring approximately half as long as the disk. A stinging spine with strong serrations, measuring up to 35 cm long and equipped with a venom gland is located about a third of the distance along the tail. A second or even third spine may also be present. Tail behind the spine bears a low cutaneous fold on top and a short, deep fold underneath. Body and tail are smooth with a few dermal denticles on the leading edge of the disk. Older individuals develop a row of bony knobs along the midline of the back. Coloration of this species is gray, brown, reddish, or olive-green above, and whitish below with dark fin margins. Young rays may have white spot. It has been reported to reach a DW of 1.5 m and a total length of 2.5 m.

Food habits: Feeds on a wide variety of bottom-dwelling organisms like crustaceans, cephalopods, bivalves, polychaete worms, and small bony fishes.

Reproduction: This species shows ovoviparity (aplacental viviparity), with embryos feeding initially on yolk. Additional nourishment for the embryos is from the mother by indirect absorption of uterine fluid enriched with mucus, fat, or protein through specialized structures. Gestation period is about 4 months. It gives birth to 4–7 pups. Newborns measure 8 cm across and 20 cm long. Males reach sexual maturity at 22–32 cm across and females at 24–38 cm across. Life span of this species is 21 years.

Predators: Not reported.

Parasites

Microsporidia: *Dasyatispora levantinae*

Monogenea: *Entobdella diadema*

Heterocotyle minima

Heterocotyle pastinacae

Neoentobdella diadema

Cestoda: *Acanthobothrium crassicolle*

Acanthobothrium intermedium

Dollfusiella aculeata

Dolifusiella spinifer

Dollfusiella tenuispinis
Echinobothrium typus
Grillotia erinaceus
Grillotia sp.
Kotorella pronosoma
Parachristianella monomegacantha
Parachristianella trygonis
Phyllobothrium pastinacae
Prochristianella papillifer
Progrillotia dasyatidis
Progrillotia pastinacae
Scalithrium minimum
Tetrarhynchobothrium striatum
IUCN conservation status: Data deficient.

3.1.31   *DASYATIS RUDIS* (Günther, 1870)

Common name: Smalltooth stingray.

Geographical distribution: Tropical; eastern Atlantic Ocean: Nigeria (Old Calabar), Benin, Gulf of Guinea.

Habitat: Marine; brackishwater; demersal; coastal species capable of migrating into freshwaters.

Distinctive features: It is a large stingray species entirely covered by diskrete prickles, but without tubercles. But on the middle of tail, somewhat larger ossifications are seen with a stellate base. Pectoral fins of this species are fused with the head, forming a diamond-shaped flat disk. Behind the eyes are spiracles. On the ventral surface of the disk, there are five pairs of gill slits, mouth, and nostrils. Teeth are arranged in a staggered manner to form a flat surface. This species grows to a maximum total length of 320 cm.

Food habits: Diet of this species includes small fish, snails, clams, shrimp, and some other small sea creatures.

Reproduction: This species shows ovoviparity (aplacental viviparity), with embryos feeding initially on yolk. Additional nourishment for the embryos is from the mother by indirect absorption of uterine fluid enriched with mucus, fat, or protein through specialized structures.

Predators: Not reported.

Parasites: Not reported.

IUCN conservation status: Data deficient.

3.1.32 *DASYATIS SABINA* (Lesueur, 1824)

Common name: Atlantic stingray.

Geographical distribution: Western Atlantic: Chesapeake Bay to southern Florida in the USA and the Gulf of Mexico.

Habitat: Marine; freshwater; brackishwater; demersal; estuaries, lagoons and rivers.

Distinctive features: Disk of this species has broadly rounded outer corners. Snout is prominent and triangular. Upper surface is brown or yellowish brown and paler toward margins of disk. Lower surface is white. Few scapular spines are present. Mid-dorsal row of spines is also present. Beyond pelvic fins, tail has few spines. Maximum size (DW) of female and male is 45 and 33 cm, respectively. This species grows to a maximum weight of 4.9 kg (Robins & Ray, 1986).

Food habits: Feeds on tube anemones, polychaete worms, small crustaceans, clams, and serpent stars.

Reproduction: This species shows ovoviparity (aplacental viviparity), with embryos feeding initially on yolk. Additional nourishment for the embryos is from the mother by indirect absorption of uterine fluid enriched with mucus, fat, or protein through specialized structures. Size at maturity for female and male is 22–25 cm (DW) and 20–21 cm (DW), respectively. Gestation period is 4 months; average litter size is 2.3–2.6 and size at birth is 10–13 cm (DW).

Predators: Whale shark, tiger shark, and bull shark are the major predators. Fresh water populations like stingrays are also preyed upon by alligators.

Parasites

Monogenea: *Thaumatocotyle roumillati*

Cestoda: *Acanthobothrium brevissime*

Dollfusiella tenuispinis

Parachristianella monomegacantha

Pterobothrium lintoni

Prochristianella hispida

Prochristianella sp.

Trematoda: *Nagmia floridensis*

Myliobaticola richardheardi

Copepoda: *Argulus* sp.

IUCN conservation status: Least concern.

3.1.33 *DASYATIS SAY* (Lesueur, 1817)

Common name: Bluntnose stingray.

Geographical distribution: Subtropical; western Atlantic: New Jersey, USA and northern Gulf of Mexico to Argentina; West Indies; Antilles.

Habitat: Marine; demersal; nearshore, estuaries and surf zones.

Distinctive features: This species has a diamond-shaped pectoral fin disk which is about one-sixth wider than long, with broadly rounded outer corners. Leading margins of the disk are nearly straight and converge at the tip of the snout. Mouth is curved, with a central projection on the upper jaw that fits into an indentation on the lower jaw. There is a row of five papillae across the floor of the mouth. There are 36–50 upper tooth rows. Teeth have quadrangular bases and are arranged with a quincunx pattern into flattened surfaces. Tooth crowns are rounded in females and in males, they are triangular and pointed. Pelvic fins are triangular with rounded tips. Whip-like tail measures over one and a half times as long as the disk and bears one or two long, serrated stinging spines on top. After the spine, there are well-developed upper and lower fin folds, with the lower fold longer and wider than the upper. Small thorns or tubercles are found in a midline row from behind the eyes to the base of the tail spine. Adults also have prickles before and after the eyes and on the outer parts of the disk. Dorsal coloration is grayish, reddish, or greenish-brown. Some individuals may also possess bluish spots which are darker toward the sides and rear or have a thin white disk margin. Belly is whitish sometimes with a dark disk margin or dark blotches. Maximum DW of this species is 1 m (Florida Museum of Natural History—Ichthyology, http://www.flmnh.ufl. edu/index.php?cID=1859).

Food habits: Feeds on small fish, clams, marine worms, and crustaceans (i.e., shrimp, crabs).

Reproduction: This species shows ovoviparity (aplacental viviparity), with embryos feeding initially on yolk. Additional nourishment for the embryos is from the mother by indirect absorption of uterine fluid enriched with mucus, fat, or protein through specialized structures. Only the left ovary and uterus in adult females are functional. Each female gives birth to 2–4 pups. Newborn rays measure 15–17 cm across and weigh 170–250 g. Males mature sexually at a DW of 30–36 cm and a weight of 3–6 kg, while females mature at a DW of 50–54 cm and a weight of 7–15 kg.

Predators: It is preyed upon by larger fishes such as the bull shark, *Carcharhinus leucas.*

Parasites

Monogenea: *Listrocephalos corona*

Thaumatocotyle longicirrus

Thaumatocotyle retorta

Cestoda: *Acanthobothrium brevissime*

Kotorella pronosoma

Prochristianella hispida

Prochristianella sp.

Pterobothrium heteracanthum

Trimacracanthus binuncus

IUCN conservation status: The World Conservation Union (IUCN) does not consider this species as a threatened or endangered species.

3.1.34   DASYATIS SINENSIS (Steindachner, 1892)

Image not available.

Common name: Chinese stingray.

Geographical distribution: Subtropical; northwest Pacific: Shanghai, China.

Habitat: Marine; demersal; cold, inshore waters.

Distinctive features: Pectoral fin disk of this species is diamond shaped and is almost as long as wide, with slightly convex leading and trailing margins. Snout is triangular and projecting with a quarter of the disk length. Eyes are of moderate size and are closely followed by a pair of spiracles. Mouth is bow shaped with five papillae on the floor. Tooth rows number 37 in the upper jaw and 40 in the lower jaw. Teeth of adult males are pointed and that of females are blunt. Whip-like tail bears both dorsal and ventral fin folds behind a stinging tail spine. Some individuals lack a tail spine. Dorsal surface is roughened by a band of small dermal denticles, extending from the snout to the base of the tail. Coloration of this species is gray above, lightening to yellowish toward the fin margins, and lighter below. It attains a DW of 40 cm and a total length of 82 cm for males and 73 cm for females.

Food habits: Little is known of the Chinese stingray's natural history.

Reproduction: This species exhibits ovoviparity (aplacental viviparity), with embryos feeding initially on yolk. Additional nourishment for the embryos is from the mother by indirect absorption of uterine fluid enriched with mucus, fat, or protein through specialized structures.

Predators: Not reported.

Parasites: Not reported.

IUCN conservation status: Data deficient.

3.1.35 *DASYATIS THETIDIS,* Ogilby, 1899

Common name: Thorntail stingray.

Geographical distribution: Subtropical; Indo-West Pacific: Mozambique, South Africa, Reunion, southern Australia and New Zealand.

Habitat: Marine; demersal; inshore waters, estuaries, lagoons and reefs; freshwaters.

Distinctive features: This species has a diamond-shaped pectoral fin disk which is about one-fourth wider than long. Mouth is slightly arched. There are five papillae across the floor. Tooth rows number 25–43 in the upper jaw and 29–48 in the lower jaw and are arranged with a quincunx pattern into pavement-like surfaces. Pelvic fins have rounded tips and gently curved trailing margins. Whip-like tail measures about twice the length of the disk and bears one or two long stinging spines with up to 88 serrations. Adults have a row of large, sharp thorns running along the midline of the back from behind the eyes to the tail spine. Thorns of various sizes are also found scattered about the dorsal surface of the disk. Tail behind the spine is also densely covered by stout thorns. Coloration of this species is a uniform dark brown or gray to black above and whitish below. It reaches 4 m long, 1.8 m across, and 214 kg in weight.

Food habits: Preys mainly upon crabs, mantis shrimp, bivalves, polychaete worms, and conger eels.

Reproduction: This species shows ovoviparity (aplacental viviparity), with embryos feeding initially on yolk. Additional nourishment for the embryos is from the mother by indirect absorption of uterine fluid enriched with mucus, fat, or protein through specialized structures.

Predators: Not reported.

Parasites

Cestoda: *Dolifusiella ocallaghani*

Kotorella pronosoma

Oncomegas favensis

Prochristianella clarkeae

Trimacracanthus aetobatidis

Nematoda: *Echinocephalus overstreeti.*

IUCN conservation status: Data deficient.

3.1.36 *DASYATIS TORTONESEI,* Capapé, 1975

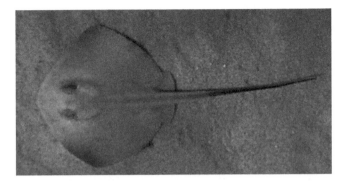

Common name: Tortonese's stingray.

Geographical distribution: Temperate; northeast Atlantic: Mediterranean Sea.

Habitat: Marine; demersal; sandy and muddy bottoms.

Distinctive features: Disk of this species is rhomboid. While front margin is more or less sinuous, hind margin is convex. Snout is obtuse, but a little produced. Its tip is projecting a little beyond front margin of disk. Tail is 1.2–1.4 times the length of disk, with a relatively short and narrow membranous fold below and a dorsal ridge on top. Floor of mouth is with three filamentous papillae. Upper surfaces are without large tubercles or bucklers on disk, but more or less covered with denticles in larger individuals. Large broad-based thorns are seen on tail. Upper surfaces are olive-brown to grayish and underside is white with a brown border around disk. It grows to a maximum size of 80 cm DW.

Food habits: Feeds on bottom-living invertebrates and fishes.

Reproduction: This species shows ovoviparity (aplacental viviparity), with embryos feeding initially on yolk. Additional nourishment for the embryos is from the mother by indirect absorption of uterine fluid enriched with mucus, fat, or protein through specialized structures. Gestation period is 4 months and 6–9 young are produced.

Predators: Not reported.

Parasites

Monogenea: *Heterocotyle capapei*

Thaumatocotyle tunisiensis

Cestoda: *Dollfusiella aculeata*

Parachristianella trygonis

Prochristianella papillifer

Progrillotia dasyatidis

IUCN conservation status: Data deficient.

3.1.37 *DASYATIS USHIEI* (Jordan & Hubbs, 1925)

Common name: Cow stingray.

Geographical distribution: Temperate; northwest Pacific: northern Japan to the East China Sea.

Habitat: Marine; demersal; continental shelves.

Distinctive features: Disk of this species is broader than long. Snout is slightly produced. Eye is somewhat elevated. Spiracles are fairly large. Mouth is little less than two thirds snout, with seven papillae, in three groups of which median comprises three papillae. Teeth are placed in 24 rows above and 27 rows below. Body is smooth and only a series of tubercles is seen on the posterior of tail. Tail is longer than twice disk length and is compressed anteriorly and whip like behind spine base. Average STL of males is 74 mm and average total number of serrations of the spine of this sex is 92. Body coloration is grayish brown above and whitish below. Maximum size of this species is 202 cm DW.

Food habits: Nothing is known (Ishihara and Valenti, 2009).

Reproduction: This species shows ovoviparity (aplacental viviparity), with embryos feeding initially on yolk. Additional nourishment for the embryos is from the mother by indirect absorption of uterine fluid enriched with mucus, fat, or protein through specialized structures.

Predators: Not reported.

Parasites: Not reported.

IUCN conservation status: Data deficient.

3.1.38   *DASYATIS ZUGEI* (Müller & Henle, 1841)

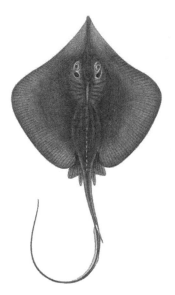

Common name: Pale-edged stingray.

Geographical distribution: Tropical; Indo-West Pacific: India to southern Japan, Myanmar, Malaya, Indonesia, China, and Indo-China.

Habitat: Marine; brackishwaters; demersal; amphidromous; estuaries.

Distinctive features: It has a diamond-shaped pectoral fin disk which is slightly longer than wide with concave anterior margins merging into an elongated, triangular snout. Head comprises more than half of the disk length. Eyes are small, with a pair of much larger spiracles immediately behind. Mouth is gently curved, without papillae on the floor. There are 40–55 tooth rows in either jaw, arranged with a quincunx pattern into pavement-like surfaces. Teeth of adult males have pointed cusps and that of females are blunt. Pelvic fins are triangular. Tail is whip like, much longer than the disk, and bears a stinging spine. Average STL in males is 53 mm and 49 mm in females, and the total number of serrations in these sexes was 54 and 64, respectively. Adults have a row of 5–9 small tubercles in front of the spine. Dorsal surface is a uniform chocolate brown, darkening on the tail fin folds, while belly is white with a brown band around the margin of the disk. This species reaches a maximum total length of 75 cm and DW of 29 cm (Schwartz, 2007).

Food habits: Feeds on bottom-dwelling organisms (small crustaceans and small fishes).

Reproduction: This species shows ovoviparity (aplacental viviparity), with embryos feeding initially on yolk. Additional nourishment for the embryos is from the mother by indirect absorption of uterine fluid enriched with mucus, fat, or protein through specialized structures. Females give birth to 1–3 young ones at a time and there is no defined reproductive seasonality. Newborns measure 8–10 cm across. Males mature sexually at a DW of 18 cm and females at a DW of 19 cm.

Predators: Not known.

Parasites

Monogenea: *Trimusculotrema schwartzi*

Cestoda: *Acanthobothrium grandiceps*

Acanthobothrium ijimai

Acanthobothrium indiana

Acanthobothrium micracantha

Acanthobothrium zugeinensis

Balanobothrium dattatrayae

Balanobothrium yamagutii

Echeneibothrium karbharae

Hexacanalis thapari

Kotorella pronosoma

Parachristianella sp.

Pedibothrium zugei

Phyllobothrium bifidum

Polypocephalus ratnagiriensis
Polypocephalus visakhapatnamensis
Rhinebothrium xiamenensis
Sephenicephalum bombayensis
Sephenicephalum dnyandevi
Tetragonocephalum madhulatae
Tetragonocephalum madrasensis
Tetragonocephalum raoi
Tylocephalum alii
Tylocephalum singhii
Uncibilocularis indiana
Uncibilocularis shindei
Uncibilocularis veravalensis

IUCN conservation status: Near threatened.

3.1.39 HIMANTURA ALCOCKII (Annandale, 1909) = *Himantura macrura*

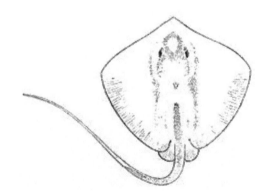

Phylum: Chordata   Subphylum: Vertebrata
Class: Chondrichthyes   Subclass: Elasmobranchii
Order: Myliobtiformes   Family: Dasyatidae

Common name: Pale-spot whip ray.

Geographical distribution: Tropical; Indian Ocean: India to Indonesia.

Habitat: Marine; brackish; benthopelagic; coastal waters and river mouths.

Distinctive features: Side of head of this species is continuous with the anterior margin of pectoral fin. Dorsal fin is totally absent or indistinct, when present. Disk is about 1.2 times as broad as long. There is no caudal fin. Tail is long and whip like. This species attains a maximum size of 100 cm DW.

Food habits: Diet presumably consists of bivalve molluscs, crustaceans, and small fishes.

Reproduction: This species shows ovoviparity (aplacental viviparity), with embryos feeding initially on yolk. Additional nourishment for the embryos is from the mother by indirect absorption of uterine fluid enriched with mucus, fat, or protein through specialized structures. Female gives birth to litters of 1–4 pups after an unknown gestation period. Males reach maturity at ~48 cm DW. Size at maturity in females is >54 cm DW. Size at birth is 18–21 cm.

Predators: Not known.

Parasites: Not reported.

IUCN conservation status: Vulnerable.

3.1.40 HIMANTURA ASTRA, Last et al., 200

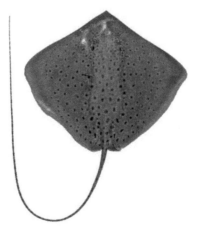

Common name: Black-spotted stingray, Blackspotted whipray, Black-spotted whipray, Coachwhip ray.

Geographical distribution: Tropical; Indo-West Pacific: Australia to West Papua, Indonesia, including Timor Sea.

Habitat: Marine; benthopelagic; coastal waters; favoring sandy habitats.

Distinctive features: It has a diamond-shaped, relatively thin pectoral fin disk which is 1.1–1.2 times wider than long. Outer corners of the disk are fairly angular, and the leading margins are nearly straight. Snout is triangular and forms a broadly obtuse angle with a barely protruding, pointed tip. Eyes are small and immediately followed by larger, oval spiracles. Mouth is strongly bow shaped and contains a row of four papillae (nipple-like structures) across the floor. There are small, blunt teeth with 41–49 rows in the upper jaw and 40–50 rows in the lower jaw. Pelvic fins are small and slightly narrow, with a curving posterior margin. Males have stout claspers. Very thin, gently tapering whip-like tail which measures 2.1–2.7 times as long as the disk is wide and lacks fin folds. One or two slender stinging spines are present atop the tail. Behind the sting, there is a deep ventral groove and prominent lateral ridges running to the tip of the tail. Upper surface of the disk is densely covered by tiny heart-shaped dermal denticles in a wide central band from between the eyes to entirely cover the tail. Coloration of this species is grayish-brown above, with many small dark spots covering all or part of the disk and tail base. Tail past the sting bears alternating light and dark saddles. Disk and tail are plain white below. It has the maximum size of 180 cm DW.

Food habits: Over 90% of the black-spotted whipray's diet consists of crustaceans (shrimps, stomatopods, and crabs). Other preferred diet is polychaete worms.

Reproduction: It is aplacental viviparous, with females nourishing their young with nutrient-rich uterine milk. Females have a single functionary ovary and uterus (on the left) and produce litters of 1–3 pups. Newborns measure 17–19 cm across and both sexes mature at 46–50 cm across. Growth rate is relatively low, with females growing much more slowly than males.

Predators: Unknown.

Parasites: Not reported.

IUCN conservation status: Least concern.

3.1.41 *HIMANTURA BLEEKERI* (Blyth, 1860) = *Himantura uarnacoides*

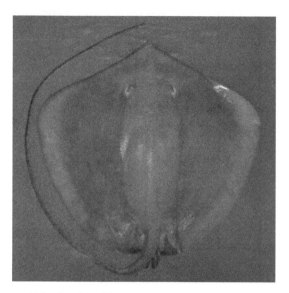

Common name: Bleeker's whipray.

Geographical distribution: Tropical; Indo-Pacific: from Pakistan to the Malay Peninsula.

Habitat: Marine; brackish; benthopelagic; amphidromous; Inshore species; soft substrates to at least 30 m; estuaries.

Distinctive features: In this species, two long finger-like processes are seen on floor of mouth. Tail is long and whip like. Large round tubercle is seen in middle of back and usually three smaller ones before and three more behind. Others are along upper surface of tail to caudal spine. There is no cutaneous fold. Coloration is uniformly dark-brown above. Ventral surface is white with broad dark-brown margin increasing with age. Maximum size of this species is 105 cm DW (Munro, 2000).

Food habits: Feeds primarily on crustaceans followed by teleosts, molluscs, and polychaetes.

Reproduction: This specie shows ovoviparity (aplacental viviparity), with embryos feeding initially on yolk. Additional nourishment for the embryos is from the mother by indirect absorption of uterine fluid enriched with mucus, fat, or protein through specialized structures.

Predators: Unknown.

Parasites

Cestoda: *Tetragonocephalum aetobatidis*

Tylocephalum choudhurai

Tylocephalum haldari

Thysanocephalum karachii

Tylocephalum krisnai

IUCN conservation status: Vulnerable.

3.1.42   HIMANTURA FAI, Jordan & Seale, 1906

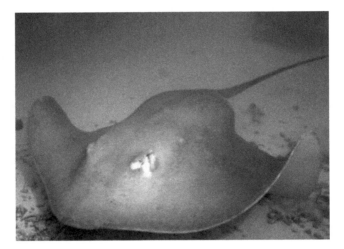

Common name: Pink whipray.

Geographical distribution: Tropical; Indo-Pacific: possibly widespread from South Africa to Micronesia.

Habitat: Marine; reef-associated; lagoon and seaward sand flats from the intertidal to at least 200 m.

Distinctive features: The pectoral fin disk of this species is diamond shaped and thick at the center, measuring 1.1–1.2 times wider than long. Outer corners of the disk are angular. Snout forms a very obtuse angle, with the tip protruding. Small, widely spaced eyes are followed by larger spiracles. Mouth is fairly small and surrounded by prominent furrows.

Lower jaw has a small indentation in the middle. There are two large central and two tiny lateral papillae on the floor of the mouth. Teeth are small and arranged into pavement-like surfaces. Five pairs of gill slits are located below the disk. Pelvic fins are small and narrow. Tail is extremely long and thin, measuring at least twice the disk length. It lacks fin folds and usually bears a single serrated stinging spine. Average STL is 184 mm and total number of serrations is 183. Among all North-Pacific stingrays, this species has the highest number of serrations in its spine. Adults have small, rounded dermal denticles covering the central dorsal surface of the disk, beginning in front of the eyes and extending to cover the entire tail. There are also small, sharp thorns on the midline, which become densest at the base of the tail. Coloration of this species is uniform grayish to brownish-pink above, becoming dark gray to black on the tail past the sting, and uniformly light below. Maximum size of this species is 183 cm DW and weight is 18.5 kg (Schwartz, 2007).

Food habits: Feeds on small fish, snails, clams, and shrimp, and some other small sea creatures.

Reproduction: Like other stingrays, this species is aplacental viviparous, with the mother supplying her developing embryos with nutrient-rich uterine milk through specialized uterine structures. Newborns measure 55–60 across. Males reach sexual maturity at 1.1–1.2 m across, while the maturation size of females is unknown.

Predators: Not reported.

Parasites

Monogenea: *Entobdella* sp.

Heterocotyle capricornensis

Merizocotyle australensis

Monocotyle helicophallus

Monocotyle spiremae

Neoentobdella parvitesticulata

Trimusculotrema heronensis

Cestoda: *Dollfusiella spinulifera*

Prochristianella spinulifera

IUCN conservation status: Least concern.

3.1.43 *HIMANTURA FAVA* (Annandale, 1909)

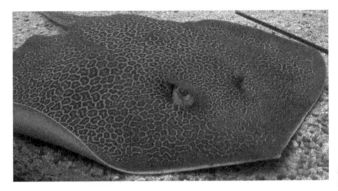

Common name: Honeycomb whipray.

Geographical distribution: Indo-Pacific: off India, Indonesia, and Gulf of Thailand.

Habitat: Inshore on soft substrates.

Distinctive features: This species has a diamond-shaped, thin pectoral fin disk which is lightly wider than long. A pointed, protruding snout is seen. Eyes are small and immediately followed by much larger spiracles. Mouth is strongly bow shaped and contains a pair of papillae (nipple-shaped structures) on the floor. Teeth are small and have a low, transverse ridge on the crown. Pelvic fins are small and roughly triangular. Tail is thin and whip like, measuring about twice as long as the disk and lacks fin folds. A single serrated, stinging spine is seen on the dorsal surface, relatively close to the tail base. Adults have a broad band of small, flattened dermal denticles running centrally from before the eyes, over the back, onto the tail. At the center of the disk, there is an enlarged, round "pearl" denticle trailed by 2–3 smaller thorns along the midline. With age, dorsal coloration becomes a honeycomb-like pattern of large, nearly black rings (ocelli) and reticulations, which are separated from each other by thin yellow lines. Tail is covered by alternating dark and light bands or other markings past the sting. Belly is white. This species has been reported to have a maximum size of 1.3 m DW (White et al., 2006).

Food habits: Diet consists of crustaceans and small fishes.

Reproduction: It is aplacental viviparous as in other stingrays, with the mother providing her developing embryos with uterine milk. Newborns measure 26–27 cm across. Males reach sexual maturity at 60–70 cm across.

Predators: Unknown.

Parasites: Cestoda: *Dollfusiella* sp.

IUCN conservation status: Not evaluated.

3.1.44 *HIMANTURA GERRARDI* (Gray, 1851)

Common name: Sharpnose stingray.

Geographical distribution: Tropical; Indo-Pacific: India to New Guinea, north to Japan. Red Sea and east African coast; River Ganges.

Habitat: Marine; brackishwaters; demersal; rivers and estuaries; favors sandy and mud bottoms.

Distinctive features: A plain light angular stingray species with a sharply pointed snout and a long slender tail which is over twice the body length. There are no caudal fin folds and large thorns. A band of small flat denticles is seen along mid-back and usually one medium-sized sting is seen on tail. Average STL of males is 23 mm and 37 mm in females. Total number of serrations in these sexes is 41 and 83, respectively. Color is light brown above and white below. Tail is with transverse bands of light and dark brown which may fade in adults. This species has a maximum total length of 2 m and DW of 90 cm (Manjaji et al., 2009; Schwartz, 2007).

Food habits: Feeds on bottom crustaceans including shrimps, crabs and small lobsters.

Reproduction: This species shows ovoviviparity (aplacental viviparity), with embryos feeding initially on yolk. Additional nourishment for the embryos is from the mother by indirect absorption of uterine fluid enriched with mucus, fat, or protein through specialized structures. It gives birth to two pups.

Predators: Not reported.

Parasites

Monogenea: *Dendromonocotyle colorni*

Dendromonocotyle lotteri

Empruthotrema dasyatidis

Merizocotyle australensis

Cestoda: *Kotorella pronosoma*

Tetrarhynchobothrium sp.

Trygonicola macroporus

IUCN conservation status: Vulnerable.

3.1.45 *HIMANTURA GRANULATA* (Macleay, 1883)

Common name: Mangrove whipray.

Geographical distribution: Tropical; Indo-West Pacific: Maldives, Seychelles, New Guinea, Micronesia, south to Australia.

Habitat: Marine; brackish; mangrove areas, also over sand or sand and rubble in lagoons near reefs to depth of 85 m on the continental shelf.

Distinctive features: The pectoral fin disk of this species is very thick and oval in shape, measuring 0.9–1.0 times as wide as long. Anterior margins of the disk are nearly straight and converge at a broad angle on the snout tip. Medium-sized, widely spaced eyes are immediately followed by the spiracles. Lower jaw is bow shaped, and there are 0–5 papillae across the floor of the mouth. Teeth are arranged in a quincunx pattern with 40–50 rows in the upper jaw and 38–50 rows in the lower jaw. There are five pairs of gill slits beneath the disk. Pelvic fins are small and narrow. Tail is thick at the base and measures 1.5–2 times longer than the DW. One or two dorsally positioned serrated stinging spines are located in the first-third of the tail. Average STL of males and females is 52 mm and total number of serrations in these sexes is 124 and 97, respectively, with 2–13 serrations occurring on the sides of spine base. Beneath the sting, tail becomes thin and whip like, without fin folds. Upper surface of the body and tail is roughened by tiny dermal denticles, which become larger toward the midline of the back and tail. In addition, one or two irregular rows of thorns are present along the dorsal midline from the head to the sting. Coloration is dark brown to gray above with many white dots and flecks. Belly is white with small dark spots toward the disk margin. Tail abruptly becomes white past the sting. This species grows to 1.4 m across and over 3.5 m long (Schwartz, 2007).

Food habits: Feeds on small fishes, bottom dwelling crustaceans, and large infauna.

Reproduction: This species shows ovoviparity (aplacental viviparity), with embryos feeding initially on yolk. Additional nourishment for the embryos is received from the mother by indirect absorption of uterine fluid enriched with mucus, fat, or protein through specialized structures. Newborns measure 14–28 cm across, and males reach sexual maturity at 55–65 cm across.

Predators: Not known.

Parasites

Monogenea: *Heterocotyle granulata*

Cestoda: Rhinebothrium himanturi

Rhinebothrium sp.

IUCN conservation status: Near threatened.

3.1.46 *HIMANTURA HORTLEI*, Last et al., 2006

Common name: Hortle's whipray.

Geographical distribution: Tropical; western Pacific: southern Irian Jaya and Papua New Guinea.

Habitat: Marine; brackish; demersal.

Distinctive features: Pectoral fin disk of this species is heart shaped and slightly longer than wide. Anterior margins are concave and converge on a highly elongated, narrowly triangular snout. Eyes are tiny and spaced wide apart and immediately followed by large, teardrop-shaped spiracles. Mouth is strongly bow-shaped and does not have papillae (nipple-shaped structures). Small, blunt teeth are set closely with a quincunx pattern and are stained orange to brown in adults. There are 21–25 upper and 24–28 lower tooth rows. Five pairs of gill slits are clearly S-shaped. Pelvic fins are short and broad. Very thin tail measures 2.6–3.4 times as long as the body and lacks fin folds. One or two stinging tail spines are placed on the upper surface of the tail. A wide band of flattened dermal denticles runs along the dorsal surface of the disk from before the eyes to the tail. Small, sharp denticles are also found scattered over the snout and concentrated at the tip. Tail past the sting is uniformly covered by denticles. Coloration of body is yellowish brown. Tail is uniformly brown and lighter in front of the sting. Belly is bright yellow, with a thin dark border around the disk margin and sometimes darker blotches around the nostrils, mouth, and gill slits. Largest known male is 71 cm across and largest female measures 65 cm across (Last et al., 2006).

Food habits: It is a predator of crustaceans, molluscs, and small fishes.

Reproduction: Reproduction in this species is aplacental viviparous, with the females supplying their developing embryos with uterine milk. Young are born at under 20 cm across.

Predators: Not known.

Parasites: Not reported.

IUCN conservation status: Vulnerable.

3.1.47   *HIMANTURA IMBRICATA* (Bloch & Schneider, 1801) = *Amphiotistius imbricatus*

Common name: Scaly whipray.

Geographical distribution: Tropical; Indo-West Pacific: Red Sea and Mauritius to Indonesia.

Habitat: Marine; freshwater; brackishwater; demersal; amphidromous.

Distinctive features: DW of this species is equal to its disk length and tail is shorter than the body. Spiracle is as large as eye. Interorbital, cranium, and middle of back are broadly covered by minute rough plates. Spines are located from middle of disk in medium line to tail. Average STL of males is 45 mm and 40 mm in females. Total number of serrations in these sexes is 76 and 75, respectively. Tail is without membranes. Dorsal coloration is reddish-brown and yellow spotted and belly is white. DW ranges of males and females are 140–340 mm and 130–360 mm, respectively. Maximum size is reported at 65 cm total length and 22 cm DW (Schwartz, 2007).

Food habits: It is a benthic carnivore feeding mostly on small crustaceans, cephalochordates (*Amphioxus* sp.), molluscs, polychaetes, and small fishes.

Reproduction: Very little is known of the biology and ecology of this species. Maturity size in female is 20.7 cm DW. Size at birth is about 10 cm DW, and males and females mature at 20.5–21 cm DW.

Predators: Not reported.

Parasites

Cestoda: *Anthobothrium septatum*

Dollfusiella sp.

Eutetrarhynchus sp.

Grillotia (Christianella) minuta

Kotorella pronosoma

Kotorella sp.

Nybelinia sp.

Parachristianella sp.

Prochristianella sp.

Pterobothrium platycephalum

Tetrarhynchobothrium rossi

Trematoda: *Orchispirium heterovitellatum*

IUCN conservation status: Data deficient.

Antimicrobial and anticoagulant activities: The spine extract of this species showed potent antibacterial activity against all tested pathogens. Maximum activity (14 mm) was found against *Staphylococcus aureus*. Crude extract showed 91.50 USP units/mg of anticoagulant activity (Kalidasan et al., 2014).

3.1.48   HIMANTURA JENKINSII (Annandale, 1909) = *Himantura draco*

Common name: Jenkin's whipray, sharpnose Stingray, brown Stingray.

Geographical distribution: Tropical; Indo-Pacific: from southern Africa and India to Australia and Papua New Guinea.

Habitat: Marine; brackish; demersal; inshore and sandy substrates.

Distinctive features: Pectoral fin disk of this species is diamond shaped and rather thick in the center, measuring 1.1–1.2 times wider than long. Outer corners of the disk are broadly rounded. Anterior margins of the disk are nearly straight and converge at a very obtuse angle on the snout, which has a protruding tip. Eyes are medium sized and closely followed by larger spiracles. Mouth is wide and slightly arched and contains four papillae (nipple-shaped structures) on the floor. Pelvic fins are small and narrow. Cylindrical, tapering tail lacks fin folds and measures slightly longer than the DW. One to three serrated, stinging spines are located atop the tail. Average STL of males is 40 mm and 34 mm in females. Total number of serrations in males and females is 69 and 73, respectively. Upper surface of the disk has a granular texture and possesses a broad central band of closely spaced, flattened heart-shaped dermal denticles. One or more rows of large, spear-like thorns also run along the dorsal midline from the center of the disk to the base of the sting. This species is uniformly yellowish brown above and disk margin and belly are white. Tail is gray past the sting. It can grow up to 1.5 m across and 3.0 m long (Schwartz, 2007).

Food habits: Small teleost fishes form a substantial portion of its diet, while crustaceans are also consumed.

Reproduction: This species is aplacental viviparous. Developing embryos thrive at first by yolk, which is later supplemented by uterine milk produced by the mother. Newborns measure 20–27 cm across, and males reach sexual maturity at 75–85 cm across.

Predators: Not reported.

Parasites

Cestoda: *Dollfusiella* sp.

Iobothrium elegans

Oncomegoides celatus

Proemotobothrium sp.

Parachristianella monomegacantha

Parachristianella baverstocki

Parachristianella indonesiensis

Dollfusiella ocallaghani

Pterobothrium platycephalum

IUCN conservation status: Least concern.

3.1.49 HIMANTURA LEOPARDA, Manjaji-Matsumoto & Last, 2008

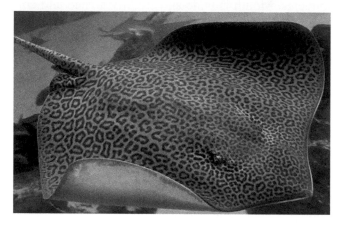

Common name: Leopard whipray.

Geographical distribution: Tropical; Indo-West Pacific: Africa to New Guinea, north to Thailand and south to Australia.

Habitat: Marine; benthopelagic; inshore and coastal waters.

Distinctive features: This species has a diamond-shaped pectoral fin disk which is wider than long and rather thick at the center, with narrowly rounded to angular outer corners. Leading margins are sinuous and converge to a broadly triangular snout. Tip of snout protrudes as a distinct, pointed lobe. Eyes are small and immediately followed by much larger, roughly rectangular spiracles. Mouth is strongly bow shaped, with shallow furrows at the corners. There are four short papillae (nipple-like structures) on the floor of the mouth. Teeth are small, conical, and blunt, numbering around 59 rows in the upper jaw. The five pairs of gill slits are S-shaped. Pelvic fins are fairly slender. Males have stout claspers. Very thin, whip-like tail measures 2.5–3.8 times as long as the disk and bears usually one serrated stinging spine. There are no fin folds. Adults have a broad band of tiny, closely spaced granules extending from before the eyes, onto to the tail. At the center of the disk, there is a midline row of up to 15 enlarged, heart-shaped denticles, with the two largest ones located one after the other between the shoulders. There are no enlarged denticles on the base of the tail. Adults are mostly covered by a leopard-like pattern of large, dark brown rings on a yellowish-brown background. Dark and light rings on the tail fade ventrally to become saddles. Belly is uniformly white. This may have two alternate color morphs which are yet to be described. Maximum length of this species is 1.1 m DW and maximum total length is 4.1 m.

Food habits: Preys on crustaceans and small fishes.

Reproduction: Like other stingrays, it is aplacental viviparous, with the developing embryos thrived by uterine milk produced by the mother. Newborns measure about 20 cm across and 92 cm long. Sexual maturity is attained at 70–80 cm across for males.

Predators: Not reported.

Parasites

Cestoda: *Parachristianella indonesiensis*

Parachristianella baverstock

Hirudinea: *Pterobdella amara*

IUCN conservation status: Vulnerable.

3.1.50   *HIMANTURA LOBISTOMA,* Manjaji-Matsumoto & Last, 2006

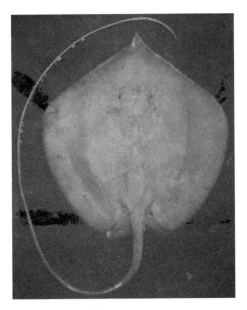

Common name: Tubemouth whipray.

Geographical distribution: Tropical; western Pacific: South China Sea, off western Borneo.

Habitat: Marine; benthopelagic; large rivers and mangrove forests, muddy bottoms.

Distinctive features: It has a diamond-shaped pectoral fin disk which is longer than wide with broadly rounded outer corners. Anterior margins of the disk are strongly concave and converge on a narrow, flattened, pointed snout. Eyes are tiny and followed by much larger, teardrop-shaped spiracles. Mouth is straight and transverse and there are no papillae (nipple-like structures) in mouth. Jaws are highly protrusible, capable of forming a tube longer than the mouth width. There are 29–34 upper and 31–36 lower tooth rows. Teeth are small, conical, and blunt and are densely arranged into pavement-like surfaces. Pelvic fins are short and can be rotated forward. Males have short, stout claspers. Slender tail measures over twice the length of the disk and lacks fin folds. A single stinging spine is seen on the upper surface of the tail near the base but is sometimes missing in adults. Upper surface of the disk and tail is covered by minute, blunt dermal denticles, with slightly larger plate-like denticles forming a distinct, broad band extending from before the eyes to the base of the tail. There are also 1–5 enlarged, oval denticles in a row between the shoulders. Coloration of this species is uniform grayish to light brown above, with the eyes and spiracles rimmed in white, and uniform white below. Females grow up to 1 m across, while males are smaller. It grows to a maximum DW of 49 cm (Manjaji-Matsumoto & Last, 2006).

Food habits: Diet of this species consists of crustaceans and small fishes.

Reproduction: Like other stingrays, it is aplacental viviparous with the developing embryos nourished by maternally produced uterine milk. Mother gives birth to one pup which measures 18 cm across. Males attain sexual maturity at 49 cm across and females at 70 cm across (Manjaji & Last, 2006).

Predators: Not reported.

Parasites: Unknown

IUCN conservation status: Vulnerable.

3.1.51 *HIMANTURA MARGINATA* (Blyth, 1860)

Common name: Blackedge whipray.

Geographical distribution: Tropical; Indian Ocean: India, Sri Lanka, and Myanmar; may venture into Indonesian waters; possibly off Mozambique.

Habitat: Marine; brackish; benthopelagic; amphidromous; inshore waters and estuaries.

Distinctive features: Pectoral fin disk of this species is diamond shaped and is wider than long and thick at the center. Outer corners are narrowly rounded and the leading margins converge at an obtuse angle. Tip of the snout projects slightly and is flanked by a pair of small, shallow concavities. Eyes are small and are followed by larger spiracles. Mouth is small and bow shaped, with a papilla (nipple-like structure) near each corner. Several teeth are arranged with a quincunx pattern. Pelvic fins extend past the disk. Tail is much longer than the disk and bears a serrated stinging spine on atop. Base of the tail is flattened and tapering to become whip like past the sting. There are no fin folds. Dorsal surface of the disk is densely covered by granules which become smaller toward the disk margin. Large thorns with star-shaped bases are seen in a row along the midline of the back, as well as scattered over the rest of the disk. Tail is smooth up to the spine, and beyond is entirely covered by small granules and prickles. Center of the disk is dark brown or gray above, becoming darker or light violet toward the margins. There may be small yellowish spots scattered over the disk, or a bluish irregular line running around the disk a small distance from the edge. Belly is white with a wide, black, irregularly edged band running along the lateral and posterior disk margins. Tail is brown at the base and white after the sting. This large species grows to 1.8 m across and is 3.5 m long.

Food habits: Feeds on small fish, snails, clams, and shrimps, and some other small sea creatures (Valenti, 2009a).

Reproduction: This species shows ovoviparity (aplacental viviparity). Embryos feed initially on yolk. Additional nourishment for the embryos is from the mother by indirect absorption of uterine fluid (enriched with mucus, fat, or protein) through specialized structures.

Predators: Not known.

Parasites

Cestoda: *Acanthobothrium dighaensis*

Hexacanalis sasoonensis

IUCN conservation status: Data deficient.

3.1.52 *HIMANTURA MICROPHTHALMA* (Chen, 1948)

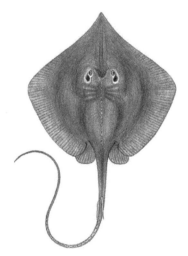

Common name: Smalleye whip ray.

Geographical distribution: Subtropical; Northwest Pacific: Taiwan.

Habitat: Marine; demersal.

Distinctive features: This species has a diamond plate body and its leading edge is slightly concave. There is a long and sharp snout. Eye is small and is slightly prominent. Eye diameter is 0.6 times the spiracle. There is a small mouth and its floor is without mastoid. Trail is slender and is 1.7 times the body length of the disk. Dorsal surface of the skin fold is completely disappeared. There is a single spine. Average total spine length is 28 mm in males and 31 mm in females. Total number of serrations in males and females is 44 and 39, respectively. Back is light brown and belly is pale (Schwartz, 2007).

Food habits: Feeds on small fish, snails, clams, and shrimp, and some other small sea creatures.

Reproduction: This specie shows ovoviparity (aplacental viviparity). Embryos feed initially on yolk. Additional nourishment for the embryos is from the mother by indirect absorption of uterine fluid (enriched with mucus, fat, or protein) through specialized structures.

Predators: Unknown

Parasites: Not reported.

IUCN conservation status: Not evaluated.

3.1.53   *HIMANTURA PACIFICA* (Beebe & Tee-Van, 1941)

Common name: Pacific chupare.

Geographical distribution: Tropical; eastern Pacific: Costa Rica and the Galapagos Islands.

Habitat: Marine; demersal; soft bottoms in shallow waters.

Distinctive features: It has a rounded pectoral fin disk and a broadly angled snout which has a small protuberance at the tip. Tail lacks fin folds but has a low ventral keel. Dorsal surface of the body and tail is covered with rough dermal denticles. There are large tubercles with four radial ridges on the shoulder region. A poisonous spine is present on the tail. This species attains a maximum length of 150 cm and a DW of 60 cm.

Food habits: Feeds on small fish, snails, clams, and shrimp, and some other small sea creatures.

Reproduction: This species shows ovoviparity (aplacental viviparity). Embryos feed initially on yolk. Additional nourishment for the embryos is from the mother by indirect absorption of uterine fluid (enriched with mucus, fat, or protein) through specialized structures.

Predators: Unknown.

Parasites

Cestoda: *Acanthobothroides pacificus*

Acanthobothroides geminium

Nematoda: *Echinocephalus janzeni*

IUCN conservation status: Not evaluated.

3.1.54 HIMANTURA PASTINACOIDES (Bleeker, 1852) = *Himantura pareh*

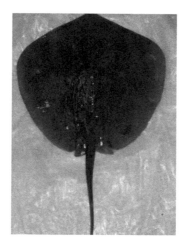

Common name: Round whip ray.

Geographical distribution: Tropical; western Central Pacific: Indonesia; Indo-Malay Archipelago including Borneo, Java and Sumatra.

Habitat: Marine; demersal; inshore on continental shelves; coastal embayments near large river mouth.

Distinctive features: Pectoral fins of this species are fused with the head, forming an oval disk. Front edge converges at an obtuse angle. Tip of snout protrudes slightly beyond the edge of the disk. Behind the eyes are spiracles. Average STL of males is 59 mm and 60 mm in females and total number of serrations in these sexes is 65 and 76, respectively. On the ventral surface of the disk, there are five pairs of gill slits, mouth, and nostrils. Small blunt teeth are arranged in a checkerboard pattern and form a flat surface. Skin folds on the caudal peduncle are absent. Tail is much greater than the length of the disk. This species reaches a maximum size of about 100 cm DW, and males mature at 43–46 cm DW (Schwartz, 2007).

Food habits: Feeds on small fish, snails, clams, and shrimp, and some other small sea creatures.

Reproduction: This species shows ovoviparity (aplacental viviparity). Embryos feed initially on yolk. Additional nourishment for the embryos is from the mother by indirect absorption of uterine fluid (enriched with mucus, fat, or protein) through specialized structures. Fecundity is said to be one pup per litter (Manjaji-Matsumoto et al., 2009).

Predators: Not reported.

Parasites

Cestoda: *Dollfusiella* sp.

Prochristianella sp.

Zygorhynchus borneensis

IUCN conservation status: Vulnerable.

3.1.55   *HIMANTURA RANDALLI,* Last et al., 2012

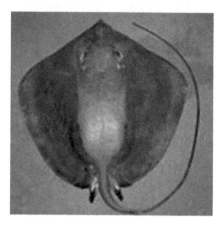

Common name: Arabian banded whipray.

Geographical distribution: Tropical; western Indian Ocean: Kuwait, Bahrain, Qatar and Iran; Persian Gulf (endemic area).

Habitat: Marine; demersal; shallow, soft substrates of mostly sand and mud.

Distinctive features: This is a medium-sized species distinguished by its weakly, rhomboidal disk. Preorbital snout is moderately elongate with weak apical lobe. There are rounded pectoral-fin apices and small, protrusible orbits. Mouth is fairly broad. Pelvic-fin base is broad. Tail behind sting is subcircular with deep longitudinal ventral groove and prominent mid-lateral ridge. One to two (usually 1) small, broadly heart-shaped to seed shaped suprascapular denticles are present. Primary denticle band and thorns are absent. Secondary denticle band is irregularly sub-oval and relatively narrow with well-defined lateral margins. Dorsal surface is uniformly colored. Dark flecks are seen pccasionally in individuals smaller than 25 cm DW. Disk margin is sometimes paler dorsally. Ventral disk is uniformly whitish. Maximum size of this species is 620 mm DW and birth size is 150–170 mm DW. Males mature at approximately 400 mm DW.

Food habits: Feeds on small fish, snails, clams, and shrimp, and some other small sea creatures.

Reproduction: This species shows ovoviparity (aplacental viviparity), with embryos feeding initially on yolk. Additional nourishment for the embryos is from the mother by indirect absorption of uterine fluid (enriched with mucus, fat, or protein) through specialized structures. Birth size is around 150–170 mm DW. Males mature at approximately 400 mm DW (Last et al., 2012).

Parasites: Not reported.

IUCN conservation status: Not evaluated.

3.1.56 *HIMANTURA SCHMARDAE* (Werner, 1904)

Common name: Chupare stingray.

Geographical distribution: Tropical; western Central Atlantic: Gulf of Campeche and the West Indies to Suriname including Brazil.

Habitat: Marine; demersal; sandy bottoms, occasionally near coral reefs.

Distinctive features: This is a large species with an oval pectoral fin disk and a long, broad-angled snout. Front margin of the disk is almost straight. Mouth is arched with indentations at the symphysis and there are five papillae on its floor. Teeth have elliptical bases and flattened cusps with a scallop-edged central depression. Upper jaw has 28–36 rows of teeth. Tail is relatively short and slender, without fin folds. There is a single saw-toothed spine which is located on the latter half of the tail. Upper surface of the body and tail is covered with small tubercles. There are large tubercles with four radiating ridges each on the shoulder region. Coloration of body is dark brown to olive above and yellowish white below, darkening to blackish toward the tip of the tail. Maximum reported DW is 2 m.

Food habits: Feeds on small fish, snails, clams, and shrimp, and some other small sea creatures.

Reproduction: This species shows ovoviparity (aplacental viviparity), with embryos feeding initially on yolk. Additional nourishment for the embryos is from the mother by indirect absorption of uterine fluid (enriched with mucus, fat, or protein) through specialized structures (Stehmann et al., 1978).

Predators: Not reported.

Parasites

Cestoda: *Acanthobothrium himanturi*

Acanthobothrium tasajerasi

Acanthobothroides thorsoni

Anindobothrium anacolum

Caulobothrium anacollum

Parachristianella cf. *monomegacantha*

Rhinebothrium magniphallum

Rhinebothrium tetralobatum

IUCN conservation status: Data deficient.

3.1.57 *HIMANTURA TOSHI,* Whitley, 1939

Common name: Black-spotted whipray.

Geographical distribution: Indo-West Pacific: Malaysia, northern Australia and New Guinea, including the Arafura Sea and Timor Sea.

Habitat: Marine; demersal; deep-waters; soft bottoms on the continental shelf.

Distinctive features: This species has a diamond-shaped and relatively thin pectoral fin disk, measuring 1.2 times wider than long. Anterior margins of the disk are nearly straight and converge on atriangular snout. Tip of the snout is pointed and slightly protruding. Eyes are moderately large and are immediately followed by spiracles. Mouth is small and bow shaped with four papillae (nipple-shaped structures) across the floor. Pelvic fins are small and narrow. Tail spine is placed on the tail's upper surface. Tail measures 2.5–3 times as long as the disk. There is a band of small, dense, heart-shaped dermal denticles extending from between the eyes to the tail. In addition, 3–4 enlarged, spear-like thorns are also present at the center of the disk, along with 1–3 preceding rows of smaller thorns. Minute denticles are also seen on the rest of the disk upper surface. Coloration of this species is uniformly dark olive-brown above, becoming lighter toward the disk margin, and uniform white below. Rarely, in larger adults, there are small pale spots or flecks near the disk margin. Tail is dark above and below, with alternating black and gray bands toward the tip. It reaches 86 cm across and is 1.7 m long.

Food habits: Feeds on prawns, small fishes, and crustaceans.

Reproduction: This species shows ovoviparity (aplacental viviparity), with embryos feeding initially on yolk. Additional nourishment for the embryos is from the mother by indirect absorption of uterine fluid (enriched with mucus, fat, or protein) through specialized structures. Females bear litters of 1–2 pups, each measuring 20–22 cm across. Males and females mature sexually at 40 cm and 66 cm across, respectively.

Predators: Not reported.

Parasites

Cestoda: *Parachristianella monomegacantha*

Prochristianella clarkeae

Zygorhynchus elongatus

IUCN conservation status: Least concern.

3.1.58 *HIMANTURA UARNAK* (Gmelin, 1789)

Common name: Honeycomb stingray.

Geographical distribution: Subtropical; Indo-Pacific: Red Sea to southern Africa and French Polynesia, north to Taiwan, south to Australia Arafura Sea and estuary of the River Ganges.

Habitat: Marine; brackish; reef-associated; amphidromous; sandy beaches and in shallow estuaries and lagoons; sandy areas of coral reefs; offshore and freshwaters.

Distinctive features: Pectoral fin disk of this species is diamond-shaped and wider than long, with the leading margins almost straight. Snout and outer corners are angular. Eyes are small and are immediately followed by a pair of spiracles. Mouth is relatively small, with a deep concavity at the center of the lower jaw and shallow furrows at the corners extending onto the lower jaw. A row of 4–5 papillae is seen across the floor of the mouth. There are 26–40 upper tooth rows and 27–44 lower tooth rows. Pelvic fins are small and triangular. Tail is whip like and is extremely thin measuring 3–3.5 times as long as the disk. Normally, one serrated stinging spine is located on the upper surface on the tail. Average STL of males is 63 mm and 55 mm in females and total number of serrations in these sexes is 92 and 73, respectively. Adults have a wide band of flattened, heart-shaped dermal denticles which extend from between the eyes to the tail spine. There are also two large thorns at the center of the back. Tail behind the spine is covered by small thorns. Adults have a dorsal pattern of several closely spaced dark brown spots or reticulations on a light brown to yellow-brown background, which becomes blackish after the spine. Belly is pale, without markings. This large species has a DW of 2 m, a total length of 6 m, and a weight of 120 kg (Schwartz, 2007).

Food habits: Feeds on small fishes, bivalves, crabs, shrimps, worms, and jellyfishes.

Reproduction: This species shows ovoviparity (aplacental viviparity), with embryos feeding initially on yolk. Additional nourishment for the embryos is from the mother by indirect absorption of uterine fluid (enriched with mucus, fat, or protein) through specialized structures. Females give birth to up to five pups after a year-long gestation period. Newborns measure 28–30 cm across and sexual maturation of females is attained at a DW of about 1 m. Males have been reported to mature at 82–84 cm across (Fricke, 1999).

Predators: Not reported.

Parasites

Monogenea: *Dendromonocotyle colorni*
Empruthotrema dasyatidis
Heterocotyle armata
Heterocotyle chinensis
Heterocotyle confusa
Heterocotyle granulatae
Monocotyle helicophallus
Monocotyle multiparous
Monocotyle spiremae
Cestoda: *Acanthobothrium cannoni*
Anthobothrium loculatum
Acanthobothrium longipedunculata
Acanthobothrium waltairensis
Dollfusiella sp.
Dollfusiella owensi
Echinobothrium reesae

Eutetrarhynchus owensi
Eutetrarhynchus sp.
Halysiorhynchus macrocephalus
Hornelliella annandalei
Kotorella pronosoma
Parachristianella baverstocki
Parachristianella indonesiensis
Parachristianella monomegacantha
Polypocephalus visakhapatnamensis
Prochristianella clarkeae
Pterobothrium sp.
Shirleyrhynchus aetobatidis
Trygonicola macroporus
Tylocephalum chiralensis
Thysanobothrium uarnakense
Uncibilocularis sidocymba
Uncibilocularis squireorum
Zygorhynchus elongatus
Zygorhynchus robertsoni
IUCN conservation status: Vulnerable.

3.1.59   HIMANTURA UNDULATA (Bleeker, 1852)

Common name: Leopard whipray.

Geographical distribution: Tropical; Indo-West Pacific: Bay of Bengal to New Guinea, north to the Ryukyu Islands, south to northern Australia.

Habitat: Marine; demersal; soft substrates, including sandy and muddy flats, lagoons; brackish estuaries and mangrove swamps.

Distinctive features: This species has a diamond-shaped, thin pectoral fin disk which is slightly wider than long; disk has broadly rounded outer corners and concave leading margins converging on a pointed, protruding snout. Eyes are small and are immediately followed by much larger spiracles. Mouth is strongly bow shaped, with shallow furrows at the corners, and contains a pair of papillae on its floor. Teeth are small and have a low, transverse ridge on the crown. Pelvic fins are small and roughly triangular. Tail is thin and whip like, measuring about twice as long as the disk, and lacks fin folds. Normally, a single serrated, stinging spine is located on the dorsal surface, relatively close to the tail base. Adults have a broad band of small, flattened dermal denticles running centrally from before the eyes, over the back and onto the tail. At the center of the disk, there is an enlarged, round denticle trailed by 2–3 smaller thorns along the midline. Dorsal coloration of this species is a honeycomb-like pattern of large, nearly black rings (ocelli) and reticulations which are separated from each other by thin yellow lines. Tail is covered by alternating dark and light bands after the sting. Belly is white. This species has a DW of 1.3 m and maximum total length of 4.1 m.

Food habits: Diet of this species is unknown but likely comprises small fishes and crustaceans (Rigby, 2012).

Reproduction: This species shows ovoviparity (aplacental viviparity), with embryos feeding initially on yolk. Additional nourishment for the embryos is from the mother by indirect absorption of uterine fluid (enriched with mucus, fat, or protein) through specialized structures. Newborns measure 26–27 cm across. Males reach sexual maturity at 60–70 cm across.

Predators: Not reported.

Parasites

Cestoda: *Kotorella pronosoma*

Tetragonocephalum passeyi

IUCN conservation status: Vulnerable.

3.1.60 HIMANTURA WALGA (Müller & Henle, 1841)

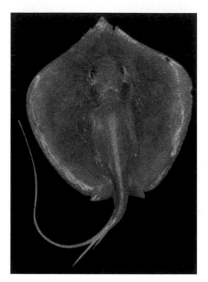

Common name: Dwarf whipray.

Geographical distribution: Tropical; western Pacific: Thailand to southeastern Indonesia.

Habitat: Marine; demersal; continental shelves and coastal embayments; muddy areas and estuaries.

Distinctive features: Disk of this species is kite shaped with pointed snout and granular dorsum. Tail is slender and whip like without markings and skin folds. Tail of female is relatively short with a bulbous end. It has 4–6 enlarged spear-like spines on the tail. Coloration of body is uniform brown above. This species grows to a maximum size of 45 cm DW.

Food habits: Feeds primarily not only on small crustaceans but also on small fishes.

Reproduction: It is an ovoviviparous species with histotrophy. Both males and females mature at 16–17 cm DW, and size at birth is 8–10 cm. Females give birth to litters of 1–2 pups (White et al., 2009).

Predators: Not reported.

Parasites

Monogenea: *Merizocotyle macrostrobus*

Cestoda: *Acanthobothrium marymichaelorum*

Echinobothrium minutamicum

Echinobothrium reesae

Eutetrarhynchus leucomelanus

Eutetrarhynchida unident

Halysiorhynchus macrocephalus

Kotorella pronosoma

Nybelinia aequidentata

Pterobothrium platycephalum

Shirleyrhynchus aetobatidis

Tetragonocephalum trygonis

Tetragonocephalum yamagutii

Trygonicola macroporus

Uncibilocularis trygonis

IUCN conservation status: Near threatened.

3.1.61   *NEOTRYGON ANNOTATA* (Last, 1987)

Phylum: Chordata   Subphylum: Vertebrata
Class: Chondrichthyes   Subclass: Elasmobranchii
Order: Myliobtiformes   Family: Dasyatidae

Common name: Plain maskray.

Geographical distribution: Eastern Indian Ocean: Timor Sea; western Pacific: Arafura Sea and northern Australia.

Habitat: Marine; demersal; continental shelf and deep waters.

Distinctive features: Pectoral fin disk of this species is thin and diamond-shaped with narrowly rounded outer corners, measuring 1.1–1.3 times longer than wide. Leading margins of the disk are gently concave and converge at a broad angle to the pointed tip of the snout. Small eyes are placed close together, and behind them are spiracles. Small mouth has prominent furrows at the corners and contains two slender papillae on the floor. Small papillae are also found around the outside of the mouth. There are five pairs of gill slits. Pelvic fins are fairly large and pointed. Tail is short and is rarely exceeding the length of the disk, and it has a broad and flattened base leading to usually two stinging spines. After the stings, tail becomes slender and bears a long ventral fin fold and a much shorter, lower dorsal fin fold. Most of the body surface lacks dermal denticles. A midline row of 4–13 small, closely spaced thorns is present behind the spiracles, and another row of 0–4 thorns before the stings. Dorsal color-ation is grayish-green and is becoming pinkish toward the disk margins. There is a dark mask-like shape around the eyes and a pair of small dark blotches is seen behind spiracles. Tail behind the stings has alternating black and white bands of variable width, ending with black at the tip. Belly is plain white and ventral fin fold is light grayish in color. This species grows to 24 cm across (DW) and 45 cm long.

Food habits: Diet consists predominantly of caridean shrimp and poly-chaete worms.

Reproduction: It is an ovoviviparous species and developing embryos are sustained by uterine milk produced by the mother. Maturing females have a single functional ovary and uterus, on the left. Litter size is one or two. Newborns measure 12–14 cm across. Males and females reach sexual maturity at a DW of 20–21 cm and 18–19 cm, respectively. The maximum life span is estimated to be 9 years for males and 13 years for females.

Predators: Not reported.

Parasites

Cestoda: *Acanthobothrium* sp.

Acanthobothrium jonesi

IUCN conservation status: Near threatened.

3.1.62 *NEOTRYGON KUHLII* (Müller & Henle, 1841) = *Dasyatis kuhlii*

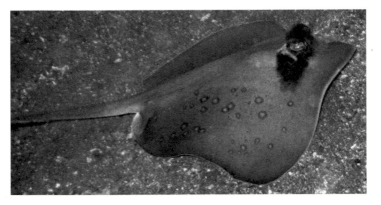

Common name: Bluespotted maskray.

Geographical distribution: Tropical; Indo-Pacific: Red Sea and East Africa to Samoa and Tonga; north to Japan; and south to Australia.

Habitat: Marine; sandy bottoms near rocky or coral reefs.

Distinctive features: This species has a flat disk-like oval body. Snout is very short and broadly angular along with its angular disk. It has a very long tail with two spines on the base of the tail. Average STL of males is 49 mm in males and 65 mm in females. Total number of spine serrations in these sexes is 60 and 73, respectively. Tail is about twice as long as the body of the ray, and barbs or spines are of two different sizes, one being very large and the other a medium-sized one. It has bright yellow eyes, and they are positioned to allow them a wide angle of view. Spiracles are located directly behind the ray's eyes. Gills and mouth are found on the underside of the body. Its coloration is dark green with blue spots with a light white underbelly. It grows to a DW of 40 cm and a total length of 70 cm (Compagno, 1986; Schwartz, 2007).

Food habits: Feeds on crabs and shrimps.

Reproduction: This species shows ovoviparity (aplacental viviparity), with embryos feeding initially on yolk. Additional nourishment for the embryos is from the mother by indirect absorption of uterine fluid (enriched with mucus, fat, or protein) through specialized structures. Mothers give birth to up to seven pups per litter. These pups range from 150 to 330 mm long at birth.

Predators: Killer whales (*Orcinus orca*), marine mammals and large fish such as sharks.

Parasites

Myxozoa: *Chloromyxum kuhlii*

Monogenea: Dendromonocotyle

Empruthotrema stenophallus

Entobdella

Cestoda: *Acanthobothrium bengalense*

Acanthobothrium confusum

Acanthobothrium pintanensis

Cephalobothrium longisegmentum

Dolifusiella michiae

Echinobothrium longicolle

Eutetrarhynchus leucomelanus

Eutetrarhynchus michiae

Halysiorhynchus macrocephaius

Mecistobothrium johnstonei

Phyllobothrium ptychocephalum

Polypocephalus kuhlii

Prochristianella aciculata

Prochristianella clarkeae

Prochristianella odonoghuei

Prochristianella omunae

Pterobothrium lintoni

Pterobothrium sp.

Tetrarhynchobothrium rossi

Trygonicola macroporus

Scalithrium trygonis

Scalithrium shipleyi

Shirleyrhynchus aetobatidis

Trematoda: Staphylorchis cymatodes

IUCN conservation status: Data deficient.

3.1.63   *NEOTRYGON LEYLANDI* (Last, 1987)

Common name: Painted maskray, brown-reticulate stingray.

Geographical distribution: Western Pacific: northern Australia and New Guinea, including the Arafura Sea and Timor Sea.

Habitat: Marine; demersal; continental shelf and deep waters.

Distinctive features: This species is known for its brown coloration and reticulate markings. Pectoral fin disk of this species is largely smooth, with a single row of thorns along the dorsal midline. Mouth is small with two central papillae and a row of enlarged, long-cusped teeth halfway along the upper jaw on both sides. Tail is very short, with well-developed dorsal and ventral fin folds and a filamentous tip, and is banded black and white past the stinging spine. It is a small maskray species, reaching a maximum size of approximately 27 cm DW.

Food habits: Diet consists of small invertebrates including crustaceans (particularly shrimps) and polychaete worms.

Reproduction: This species shows ovoviparity (aplacental viviparity), with embryos feeding initially on yolk. Additional nourishment for the embryos is from the mother by indirect absorption of uterine fluid enriched with mucus, fat, or protein through specialized structures. Litters size is 1–3 pups. Size at birth is 11 cm DW.

Predators: Not reported.

Parasites

Cestoda: *Dolifusiella owensi*

Eutetrarhynchus owensi

Trigonolobium spinuliferum

Zygorhynchus robertsoni

IUCN conservation status: Least concern.

3.1.64   NEOTRYGON NINGALOOENSIS, Last et al., 2010

Common name: Ningaloo maskray.

Geographical distribution: Subtropical; Indian Ocean: Western Australia.

Habitat: Marine; demersal; fine reddish sand close to coral reefs in inshore waters.

Distinctive features: This species has a diamond-shaped pectoral fin disk which is about 1.1 times wider than long, with straight to slightly convex leading margins and rounded outer corners. Snout is short and rounded. Eyes are large and protruding, with large crescent-shaped spiracles behind. Small mouth has shallow grooves at the corners and is surrounded by papillae. There are also two long papillae on the floor of the mouth. Teeth range from long and pointed to short and blunt. There are five pairs of S-shaped gill slits. Pelvic fins are narrow and triangular. Whip-like tail bears two very slender stinging spines on the upper surface. It is fairly broad and flattened at the base and tapers evenly past the stings. Behind the stings are well-developed dorsal and ventral fin folds. Dorsal fold is smaller than the ventral fold, which is deep and short relative to the tail's total length. Only four or five small dermal denticles (closely spaced thorns) are seen in a midline row behind the spiracles. Upper surface of the disk is yellowish brown, deepening in color toward the margins, with numerous dark orange and light blue spots. These orange spots are smaller, more sharply defined, and densest at the center of the disk, whereas the blue spots are larger, less well defined, and evenly distributed over the disk. There is a darker mask-like marking across the eyes. Tail has alternating black and white bands behind the stings. The underside is pale. This species reaches a DW of 30 cm.

IUCN conservation status: Data deficient.

Nothing much is known about its biology.

3.1.65   NEOTRYGON PICTA, Last & White, 2008

Common name: Peppered maskray, speckled maskray.

Geographical distribution: Tropical; Indo-West Pacific: Australia.

Habitat: Marine; benthopelagic; inner continental shelf; favors habitats with sandy or other fine substrates.

Distinctive features: This species has a thin, diamond-shaped pectoral fin disk which is about 1.2 times wider than long, with slightly concave leading margins and narrowly rounded outer corners. Snout forms an obtuse angle and has a pointed tip. Small, closely spaced eyes are followed by crescent-shaped spiracles. Small mouth is surrounded by papillae and bears prominent furrows at the corners. There are two papillae on the floor of the mouth. Teeth number 33–38 rows in the upper jaw and 31–40 rows in the lower; teeth are small and vary from pointed to blunt. There are five pairs of gill slits which are S-shaped. Pelvic fins are medium-sized and triangular with angular corners. Whip-like tail measures 1.0–1.3 times as long as the disk and bears two slender stinging spines on the upper surface. Tail is moderately broad and flattened at the base, becoming very thin behind the sting. Both upper and lower fin folds are present after the sting, with the upper fold shorter than the lower. There are up to 22 small, closely spaced thorns along the midline of the back behind the spiracles. Otherwise, skin is mostly smooth. This species is light yellow to brown above, with a darker reticulated pattern which may vary from faint to obvious, all overlaid by numerous black spots. There is a dark marking across the eyes resembling a mask. Tail has a pattern of saddles or bands behind the sting and its tip is white. Ventral fin fold darkens to almost black posteriorly. Belly is plain white. This species reaches 32 cm DW.

Food habits: Caridean shrimp is the most important food source. It also consumes polychaete worms and amphipods and rarely penaeid prawns, molluscs, and small bony fishes.

Reproduction: Reproduction in this species is ovoviviparous. Like other stingrays, developing embryos are initially nourished by yolk and later by uterine milk provided by the mother. Maturing females have a single functional ovary and uterus, on the left side. Females give birth to litters of one to three pups. Newborns are 9–11 cm across. Males and females reach sexual maturity at 17 and 18 cm across, respectively. Maximum life span is 11 years for males and 18 years for females.

Predators: Not reported.

Parasites: Not reported.

IUCN conservation status: Least concern.

3.1.66 *PTEROPLATYTRYGON VIOLACEA* (Bonaparte, 1832)

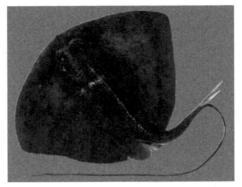

Phylum: Chordata   Subphylum: Vertebrata
Class: Chondrichthyes   Subclass: Elasmobranchii
Order: Myliobtiformes   Family: Dasyatidae

Common name: Pelagic stingray.

Geographical distribution: Subtropical; cosmopolitan in tropical and subtropical seas; eastern Atlantic: southeastern coasts of the Mediterranean and off Sicily. Eastern Pacific: California (USA), Baja California (Mexico), and Galapagos Islands; Vancouver, Chile; western Atlantic; southern Africa.

Habitat: Marine; pelagic–oceanic; open ocean waters and inshore bays.

Distinctive features: This species has a broad wedge-shaped disk which is wider than long. Snout is broadly rounded with a terminal lobe. Small eyes do not protrude from the body. Tail doubles the length of the body with a long lower caudal fin fold which terminates in front of the tip of the tail. There is no upper fin fold. Front margin of the pelvic fin is straight while the outer corner is broadly rounded. Tail has a thick base, tapering to the origin of the serrated spine. There is one or more spines located about a third of the way down the tail which are largely used for defense. Average STL of males is 109 mm and 121 mm in females with 3–20 serrations occurring on each side of spine base. Dorsal surface is dark purple or blue-green while the underside is purplish to gray. There are no distinguishing markings. It is relatively small ray reaching a maximum size of only 80 cm DW and 160 cm total length with a maximum weight of 50 kg (Schwartz, 2007, https://www. flmnh.ufl.edu/fish/discover/species-profiles/pteroplatytrygon-violacea/)

Food habits: Diet consists primarily of planktonic crustaceans and euphausiids and amphipods. Other food items include jellyfish, squid, octopus, shrimp, and small pelagic fishes such as herring and mackerel.

Reproduction: This species shows ovoviparity (aplacental viviparity), with embryos feeding initially on yolk. Additional nourishment for the embryos is from the mother by indirect absorption of uterine fluid (enriched with mucus, fat, or protein) through specialized structures. Male pelagic stingrays reach sexual maturity at 13–16 in. (35–40 cm) DW and females at 16–20 in. (40–50 cm) DW. After a gestation period of 2–4 months, female gives birth to 4–9 young with each pup measuring 15–25 cm DW (https://www.flmnh. ufl.edu/fish/discover/species-profiles/pteroplatytrygon-violacea/).

Predators: Not reported.

Parasites

Cestoda: *Acanthobothrium benedeni*

Acanthobothrium magnum

Nybelinia lingualis

Progrillotia louiseuzeti

IUCN conservation status: Least concern.

3.1.67 PASTINACHUS ATRUS (Macleay, 1883)

Phylum: Chordata   Subphylum: Vertebrata
Class: Chondrichthyes   Subclass: Elasmobranchii
Order: Myliobtiformes   Family: Dasyatidae

Common name: Cowtail stingray.

Geographical distribution: Tropical; Indo-West Pacific: Madagascar, Western Australia, Philippines, Indonesia and Malaysia, and Papua New Guinea.

Habitat: Marine; brackish; reef-associated; inshore and may even venture far upstream in estuaries.

Distinctive features: It is a large, thick, uniformly dark-colored species. Disk is rhomboidal with a wide central band of short denticles running from near snout to tail base. Tail is broad at base, tapering gradually to a sting. A wide skin fold is seen under the tail, terminating abruptly about two sting-lengths behind sting tip. This species grows to a maximum total length of 183 cm (Lizard Island Field Guide, http://lifg.australianmuseum. net.au/Group.html?groupId=eObQN8ee&hierarchyId=PVWrQCLG).

Food habits: It feeds on sandflat-associated species.

Reproduction: Reproduction in this species is ovoviviparous. Like other stingrays, developing embryos are initially nourished by yolk and later by uterine milk provided by the mother.

Predators: Not reported.

Parasites:

Cestoda: *Acanthobothrium nanogravidum*

IUCN conservation status: Not evaluated.

3.1.68 PASTINACHUS GRACILICAUDUS, Last & Manjaji-Matsumoto, 2010

Common name: Narrow tail stingray.

Geographical distribution: Tropical; western Pacific: Malaysian and Indonesian Borneo, including Sabah, Sarawak, and western and north-eastern Kalimantan.

Habitat: Marine; benthopelagic.

Distinctive features: This medium-sized species has a snout which is rounded and not produced. Apex is largely naked without enlarged denticles. Tail is compressed above mid-base of ventral cutaneous fold and its width is 0.5–0.8 times its height. Ventral fold is low and slender. There are two large, midscapular pearl thorns which are usually preceded by a smaller irregular thorn. This medium-sized species attain a maximum size of 75 cm DW and weight of 12 kg (Last & Manjaji-Matsumoto, http://bionames.org/bionames-archive/issn/1833-2331/32/115.pdf).

Food habits: Diet consists of small invertebrates including crustaceans and polychaete worms.

Reproduction: Reproduction in this species is ovoviviparous. Like other stingrays, developing embryos are initially nourished by yolk and later by uterine milk provided by the mother.

Predators: Not reported.

Parasites: Not reported.

IUCN conservation status: Not evaluated.

3.1.69 PASTINACHUS SEPHEN (Forsskål, 1775) = *Hypolophus sephen*

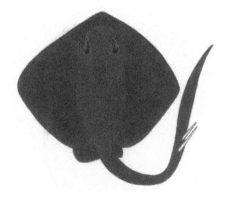

Common name: Cowtail stingray.

Geographical distribution: Tropical; Indo-West Pacific: Red Sea, Persian Gulf and South Africa to Micronesia, north to Japan, south to Melanesia and the Arafura Sea.

Habitat: Marine; freshwater; brackishwaters; reef-associated; amphidromous; rivers far from the sea.

Distinctive features: Pectoral fin disk of this species is very thick with almost straight anterior margins and rounded apices. Snout is broadly rounded and blunt. Eyes are very small and are widely spaced. Mouth is narrow, with 20 rows of distinctive hexagonal, high-crowned teeth in each jaw and five papillae on the mouth floor. Tail is broad based, with a filamentous tip and a single spine is located well backward of the pelvic fins. Average STL of males is 157 mm and 159 mm in females. Average number of serrations in both the sexes is 124. There is no upper tail fold and the high ventral tail fold measures 2–3 times the height of the tail. Disk surface is covered by a broad band of fine dermal denticles extending from near the tip of the snout to the upper surface of the tail excluding the extreme margins of the disk. Coloration of this species is a uniform grayish-brown to black above and mostly white below. Tail fold and tip are black. This species may reach 3 m long, 1.8 across (DW), and 250 kg in weight (Schwartz, 2007).

Food habits: Feeds on bony fishes (including leiognathids, *Nemipterus*, and soles), crustaceans, polychaete worms, sipunculids, and molluscs.

Reproduction: This species shows ovoviparity (aplacental viviparity), with embryos feeding initially on yolk. Additional nourishment for the embryos is from the mother by indirect absorption of uterine fluid (enriched with mucus, fat, or protein) through specialized structures. Females give birth to live young measuring 18 cm across or more.

Predators: Not reported.

Parasites

Monogenea: *Dendromonocotyle ardea*

Entobdella sp.

Heterocotyle elliptica

Neoentobdella natans

Cestoda: *Acanthobothrium bengalense*

Acanthobothrium chisholmae

Acanthobothrium gasseri

Acanthobothrium guptai
Acanthobothrium laurenbrownae
Acanthobothrium manteri
Acanthobothrium nanogravidum
Acanthobothrium semnovesiculum
Acanthobothrium walkeri
Anthemobothrium pulchrum
Balanobothrium shamraoi
Carpobothrium shindei
Cephalobothrium alii
Cephalobothrium singhi
Cephalobothrium trygoni
Dollfusiella michiae
Dollfusiella sp.
Echinobothrium deeghai
Echeneibothrium smitii
Echinobothrium reginae
Echeneibothrium trygoni
Eniochobothrium trygonis
Eutetrarhynchus leucomelanus
Eutetrarhynchus michiae
Eutetrarhynchus platycephali
Flapocephalus saurashtri
Flapocephalus trygonis
Halysiorhynchus macrocephalus
Hexacanalis indirajii
Kotorella pronosoma
Kowsalyabothrium sepheni
Lecanicephalum maharashtrae
Lecanicephalum ratnagiriensis
Mecistobothrium johnstonei
Mecistobothrium obese
Mixophyllobothrium okamuri

Oncodiscus sauridae
Oncomegas trimegacanthus
Parachristianella baverstocki
Parachristianella indonesiensis
Parachristianella monomegacantha
Phyllobothrium trygoni
Pithophorus trygoni
Polypocephalus maharashtra
Polypocephalus pratibhai
Polypocephalus thapari
Prochristianella butlerae
Prochristianella macracantha
Prochristianella odonoghuei
Prochristianella sp.
Sephenicephalum maharashtrii
Shirleyrhynchus aetobatidis
Tetragonocephalum alii
Tetragonocephalum bhagawatii
Tetragonocephalum sephensis
Tetragonocephalum shipleyi
Tetrarhynchobothrium sp.
Trygonicephalum ratnagiriensis
Tylocephalum bombayensis
Tylocephalum madhukarii
Trygonicola macroporus
Uncibilocularis bombayensis
Uncibilocularis loreni
Uncibilocularis okei
Uncibilocularis thapari
Uncibilocularis trygonis
Hirudinea: *Pterobdella amara*
IUCN conservation status: Data deficient.

3.1.70 *PASTINACHUS SOLOCIROSTRIS,* Last et al., 2005

Common name: Roughnose stingray.

Geographical distribution: Tropical; western Pacific: Malaysian Borneo and Indonesia.

Habitat: Marine; brackish; demersal; primarily in estuaries and turbid coastal marine habitats.

Distinctive features: This species has a smaller adult size with more elongate disk and head. Snout is long and more acute and its apex is covered with enlarged denticles. Sting is more posteriorly located and longer. Ventral cutaneous fold is more slender. There are enlarged pearl-shaped nuchal thorns, and fewer pectoral-fin radials and vertebrae. It grows to a maximum size of 45.0 cm DW.

Food habits: Feeds on crustaceans, polychaete worms, sipunculids, and molluscs.

Reproduction: Females mature at 50–60 cm DW and males mature at 28–40 cm DW. Reproduction is presumably viviparous, with histotrophy. Size at birth is ~22–23 cm DW. Female gives birth to one pup (Fahmi et al., 2009; Last et al., 2005).

Predators: Not reported.

Parasites

Monogenea: *Merizocotyle papillae*

Cestoda: *Echinobothrium nataliae*

IUCN conservation status: Endangered.

3.1.71 *PASTINACHUS STELLUROSTRIS,* Last et al., 2010

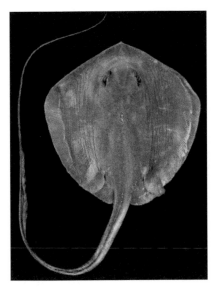

Common name: Starrynose stingray

Geographical distribution: Tropical; Asia: Indonesia, known from West Kalimantan in the vicinity of the Kapuas River estuary and near Pemangkat.

Habitat: Brackishwater, freshwater; demersal.

Distinctive features: Snout of this species is relatively long. Denticles are with broad stellate crowns. Short pungent thorns are seen along midline of tail before sting. Nuchal thorns are pearl shaped. Tail is compressed above midbase of ventral cutaneous fold. Maximum size of this species is 42.8 cm DW (Last et al., http://bionames.org/bionames-archive/issn/1833-2331/32/129.pdf).

Food habits: Little is known about its diet. It feeds primarily on small crustaceans.

Reproduction: This species shows ovoviparity (aplacental viviparity), with embryos feeding initially on yolk. Additional nourishment for the embryos is from the mother by indirect absorption of uterine fluid (enriched with mucus, fat, or protein) through specialized structures.

Predators: Not reported.

Parasites: Not reported.

IUCN conservation status: Not evaluated.

3.1.72   TAENIURA GRABATA (Geoffroy Saint-Hilaire, 1817)

Phylum: Chordata   Subphylum: Vertebrata
Class: Chondrichthyes   Subclass: Elasmobranchii
Order: Myliobtiformes   Family: Dasyatidae

Common name: Round stingray.

Geographical distribution: Subtropical; Eastern Atlantic: Canary Islands and Mauritania to Angola, including Cape Verde; Mediterranean Sea; from Tunisia to Egypt.

Habitat: Marine; demersal; neritic, coastal species; sand and rock-sand bottoms.

Distinctive features: Disk of this species is almost circular but is slightly broader than long. Tail is short and thick with a sting. It is compressed and not whip like. It has a membranous fold below. Upper surfaces are almost naked except for denticles along midline and from spiracles to origin of tail spine. Three thorns are seen on each side opposite fifth gill slits. Color of this species is gray, brown or olivaceous, with darker blotches or vermiculations. Underside is yellowish-white. This species grows to a maximum size of 250 cm total length, 100 cm DW, and weight of 84 kg.

Food habits: Feeds on bottom-living fishes and crustaceans.

Reproduction: This species shows ovoviparity (aplacental viviparity), with embryos feeding initially on yolk. Additional nourishment for the embryos is from the mother by indirect absorption of uterine fluid (enriched with mucus, fat, or protein) through specialized structures.

Predators: Not reported.

Parasites

Monogenea: *Dendromonocotyle taeniurae*

Neoentobdella apiocolpos

Cestoda: *Pterobothrium senegalense*

IUCN conservation status: Data deficient.

3.1.73 TAENIURA LYMMA (Forsskål, 1775)

Common name: Bluespotted ribbontail ray.

Geographical distribution: Tropical; Indo-West Pacific: Red Sea and East Africa to the Solomon Islands; north to southern Japan; south to northern Australia.

Habitat: Marine; reef-associated; coral reefs.

Distinctive features: This species has an oval and elongated disk with broadly rounded outer corners. Snout is bluntly rounded with slender narrow nostrils and the spiracles are large and are located close to the large eyes. Mouth and gills are located on the ventral surface of the body. Pelvic fins are moderate in size and slender. Tail is stout and more slender measuring less than twice the body length. Lower caudal fin fold is broad and reaches the tip of the tail. There are usually two, but sometimes one, medium-sized spines that are present on the tail. Average STL of males is 58 mm and 37 mm in females. Total number of serrations in these sexes is 68 and 69, respectively. Coloration is gray-brown to yellow to olive-green or reddish brown with large bright blue spots across the dorsal surface of the disk. There are blue stripes on either side of the tail. Ventral surface is uniformly white. Maximum reported size of this species is 35 cm DW and a maximum total length of 70 cm.

Food habits: Feeds on mollusks, worms, shrimps, crabs, and small fishes.

Reproduction: This species shows ovoviparity (aplacental viviparity), with embryos feeding initially on yolk. Additional nourishment for the embryos is from the mother by indirect absorption of uterine fluid (enriched with mucus, fat, or protein) through specialized structures. Females give birth to up to seven young per litter after a gestation period of 4–12 months. These young possess markings similar to that of adults including the characteristic blue spot (Randall et al., 1990).

Predators: Not reported.

Parasites

Monogenea: *Empruthotrema quindecima*

Entobdella australis

Neoentobdella australis

Pseudohexabothrium taeniurae

Cestoda: *Anthobothrium taeniuri*

Aberrapex manjajiae

Cephalobothrium taeniurai

Echinobothrium elegans

Echinobothrium helmymohamedi

Echinobothrium heroniense

Kotorella sp.

Kotoreltiella jonesi

Mecistobothrium pauciortesticulatum

Parachristianella indonesiensis

Polypocephalus saoudi

Prochristianella macracantha

Rhinebothrium ghardaguensis

Rhinebothrium taeniuri

IUCN conservation status: Near threatened.

3.1.74   TAENIUROPS MEYENI (Müller & Henle, 1841) = *Taeniura melanopilos*

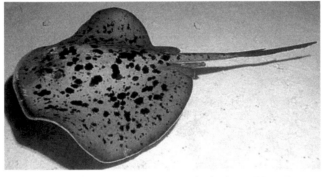

Phylum: Chordata   Subphylum: Vertebrata
Class: Chondrichthyes   Subclass: Elasmobranchii
Order: Myliobtiformes   Family: Dasyatidae

Common name: Round ribbontail ray

Geographical distribution: Tropical; Indo-West Pacific: Red Sea and East Africa to southern Japan, Micronesia; Australia and Lord Howe Island; Eastern Pacific: oceanic islands (Cocos and Galapagos).

Habitat: Marine; reef-associated; from shallow lagoons to outer reef slopes.

Distinctive features: Body of this large species is flattened and disk shaped, with the pectoral fins broadly expanded and connected to the head and body. Tail is distinctly demarcated from the disk-like body, relatively narrow, and about as long as the body length. It has a circular-shaped disk that has a mottled pattern of black, gray and white spots and blotches on its upper surface. Tail is uniformly black past the sting. Belly is pale, while the edges of the body disk and under-surface of the tail are a grayish-brown to black. Average STL of males is 60 mm and 61 mm in females. Total number of serrations in these sexes is 111 and 65, respectively. A deep and prominent skin fold runs along the underside of the tail. This species has the maximum size of 330 cm total length, 180 cm DW, and weight of 150 kg (Schwartz, 2007).

Food habits: Feeds on bottom fish, bivalves, crabs, and shrimps.

Reproduction: This species shows ovoviparity (aplacental viviparity), with embryos feeding initially on yolk. Additional nourishment for the embryos is from the mother by indirect absorption of uterine fluid (enriched with mucus, fat, or protein) through specialized structures. Female gives birth to 1–7 live young.

Predators: Potential predators of stingrays include marine mammals and large fish such as sharks.

Parasites

Monogenea: *Dendromonocotyle pipinna*

Neoentobdella garneri

Neoentobdella taiwanensis

Cestoda: *Pterobothrium minimum*

Tetrarhynchobothrium unionifactor

Nematoda: Echinocephalus sinensis

Terranova scoliodontis

Copepoda: *Eudactylina dasyati*

Eudactylina taeniuri

IUCN conservation status: Vulnerable.

3.1.75 UROGYMNUS ASPERRIMUS (Bloch & Schneider, 1801) = *Urogymnosus africanus*

Phylum: Chordata   Subphylum: Vertebrata
Class: Chondrichthyes   Subclass: Elasmobranchii
Order: Myliobtiformes   Family: Dasyatidae

Common name: Porcupine ray.

Geographical distribution: Tropical; Indo-Pacific: Red Sea and coast of East Africa to the Marshall Islands and Fiji; south to northern Australia; Eastern Atlantic: Senegal, Guinea, and Côte d'Ivoire.

Habitat: Marine; brackish; reef-associated; continental shelf area; sand and coral rubble areas near reefs; often in caves.

Distinctive features: This species has been named after its unusual thorny projections that are found on the upper side of its body. It has a thick, oval, disk-shaped body, a rounded snout and a long, whip-like tail, which does not possess the venomous barb. Instead, this species has an armored body, which is covered with a mixture of large, sharp, conical thorns and smaller, pointed projections known as denticles. Young lack the thorn but bear numerous large, flat denticles on the upper surface of the body. Coloration of this species is brown to light gray above and white below, while the tail is blackish, becoming darker toward the tip. This species grows to a maximum size of 147 cm total length and 100 cm DW.

Food habits: Forages on and around the sea-bed for bottom-dwelling crustaceans, polychaete worms, and fish.

Reproduction: This species shows ovoviparity (aplacental viviparity), with embryos feeding initially on yolk. Additional nourishment for the embryos is from the mother by indirect absorption of uterine fluid (enriched with mucus, fat, or protein) through specialized structures. Female gives birth to a litter of live young (Randall et al., 1990).

Predators: Not reported.

Parasites

Monogenea: *Dendromonocotyle pipinna*
Dendromonocotyle urogymni
Neoentobdella baggioi
Cestoda: *Acanthobothrium macracanthum*
Acanthobothrium urogymni
Rhinebothrium devaneyi
Nematoda: *Echinocephalus overstreeti*
Hirudinea: *Pterobdella amara*
IUCN conservation status: Vulnerable.

3.2   BUTTERFLY RAYS (GYMNURIDAE)

3.2.1   GYMNURA AFUERAE (Hildebrand, 1946)

Phylum: Chordata   Subphylum: Vertebrata
Class: Chondrichthyes   Subclass: Elasmobranchii
Order: Myliobtiformes   Family: Gymnuridae

Common name: Not designated.

Geographical distribution: Tropical and warm temperate; from Point Conception, southern California (United States) to Point Lobos de Afuera, Peru.

Habitat: Marine; demersal; nearshore from intertidal regions; sandy bays and beaches; silty or muddy tidal channels and estuaries.

Distinctive features: This species has a flattened body in dorsoventral direction. Females attain greater sizes than males, with a maximum reported DW of 150 cm. It feeds mainly on shellfish.

Food habits: Feeds exclusively on teleost fishes.

Reproduction: Like other gymnurids, this species bears live young but lacks a placental attachment to its embryos (aplacental viviparity). Embryos are nourished by small yolk-sacs and protein-rich uterine secretions delivered by specialized villi. Both ovaries are functional. Reported litter size ranges from 4 to 16 pups per female. Size at birth ranges from 21 to 26 cm DW. Size at first maturity is 62 cm DW for females and 41 cm DW for males.

Predators: Not known.

Parasites

Cestoda: *Acanthobothrium atahualpai*

IUCN conservation status: Not known.

3.2.2 *GYMNURA ALTAVELA* (Linnaeus, 1758)

Common name: Spiny butterfly ray, giant butterfly ray.

Geographical distribution: Subtropical; western Atlantic: southern New England, USA, Brazil to Argentina; Eastern Atlantic: Portugal to Ambriz, Angola (including the Mediterranean, Black Sea, and the Madeira and Canary islands).

Habitat: Marine; brackishwaters; demersal; sand and mud.

Distinctive features: This species has a very broad, lozenge-shaped pectoral fin disk which is much wider than long, with concave front margins and abruptly rounded corners. Snout is short and blunt. Teeth have high, conical cusps, numbering 98–138 rows in the upper jaw and 78–110 rows in the lower jaw. There is a tentacle-like structure on the inner posterior margin of each spiracle. Tail is short and slender, measuring a quarter the DW, with upper and lower fin folds. There are one or more serrated spines at the base of the tail. Skin is naked in subadults, while adults develop a patch of denticles on the center of the disk. Coloration is dark brown above, sometimes with small lighter or darker spots and blotches in a marbled pattern, and white below. Values of maximum size and weight are 4 m DW and 60 kg, respectively.

Food habits: Feeds on plankton, crustaceans, teleosts, cephalopods, lamelli branches, and gastropods.

Reproduction: This species shows ovoviparity (aplacental viviparity), with embryos feeding initially on yolk. Additional nourishment for the embryos is from the mother by indirect absorption of uterine fluid enriched with mucus, fat, or protein through specialized structures. Males mature at 78- 102 cm across and females at 108–155 cm. Gestation lasts about 6 months. Litter size is up to 8. Females have one functional ovary (the left) and two functional uteruses, with the embryos evenly distributed in each. Four to seven embryos are produced per female. Newborns measure 38–44 cm across (Bauchot, 1987).

Predators: Great hammerhead, *Sphyrna mokarran*, and marine mammals.

Parasites

Cestoda: *Anthobothrium altavelae*

Pterobothrium heteracanthum

Pterobothroides petterae

Pterobothrium lintoni

Heteronchocotyle gymnurae

IUCN conservation status: Vulnerable.

3.2.3   GYMNURA AUSTRALIS (Ramsay & Ogilby, 1886)

Common name: Australian butterfly ray.

Geographical distribution: Tropical; Indo-West Pacific: northern Australia and south coast of New Guinea; Arafura Sea.

Habitat: Marine; demersal; intertidal zones; shallow coastal waters.

Distinctive features: This species has a broad triangular wings and very short rat-like tail. This species reaches a maximum size of 73 cm DW (Allen, 2009; White, 2009).

Food habits: Feeds mainly on teleost fishes.

Reproduction: This species shows ovoviparity (aplacental viviparity), with embryos feeding initially on yolk. Additional nourishment for the embryos is from the mother by indirect absorption of uterine fluid (enriched with mucus, fat, or protein) through specialized structures. While females mature at 446 mm DW with a single functional ovary and two functional uteri, males mature at 377 mm (DW) with a single functional testis.

Predators: Not reported.

Parasites

Cestoda: *Acanthobothrium cribbi*

Nybelinia sp.

Parotobothrium balli

Tylocephalum sp.

IUCN conservation status: Least concern.

3.2.4   GYMNURA BIMACULATA (Norman, 1925)

Common name: Twin-spot butterfly ray.

Geographical distribution: Subtropical; Northwest Pacific: Yenting, Chekiang Province, China.

Habitat: Marine; demersal; sandy or muddy bottoms in inshore shallow waters.

Distinctive features: DW of this species is 2.2 times larger than the disk length. Snout length is 0.2 times the disk length. Tail is somewhat longer and its length is 0.6 times the disk length. One spine and seven black bands are seen on the tail. Average STL of females is 15 mm and average number of spine serrations in this sex is 31. Bands proximal to the spine are merged, and the two bands distal from spine are observed on a ring. There is a pair of oval bluish white ocelli at posterior side of spiracle. Other characteristics of this species are more or less similar to that of the Japanese butterfly ray. This species has a maximum total length of 25.7 cm (Isouchi, 1977; Schwartz, 2007; Shen et al., 2012).

Food habits: Feeds mainly on teleost fishes.

Reproduction: This species shows ovoviparity (aplacental viviparity), with embryos feeding initially on yolk. Additional nourishment for the embryos is from the mother by indirect absorption of uterine fluid (enriched with mucus, fat, or protein) through specialized structures.

Predators: Not known.

Parasites

Cestoda: *Anthobothrium pteroplateae*

Copepoda: *Eudactylina gymnuri*

IUCN conservation status: Data deficient.

3.2.5 GYMNURA CREBRIPUNCTATA (Peters, 1869)

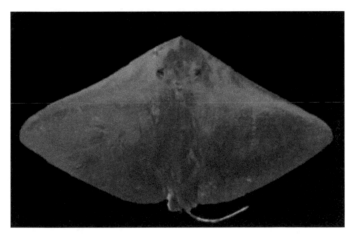

Common name: California butterfly ray, long snout butterfly ray.

Geographical distribution: Tropical; eastern central Pacific: Gulf of California and Panama.

Habitat: Marine; demersal; coastal waters, estuaries, and lagoons and occasionally in greater depths.

Distinctive features: Body of this species is flattened and is surrounded by an extremely broad disk formed by the pectoral fins, which merge in front of the head. A very short, thread-like, tail is present. It grows to a maximum total length of 31 cm.

Food habits: Feeds mainly on teleost (Bizzarro & Smith, 2012).

Reproduction: This species shows ovoviparity (aplacental viviparity), with embryos feeding initially on yolk. Additional nourishment for the embryos is from the mother by indirect absorption of uterine fluid (enriched with mucus, fat, or protein) through specialized structures.

Predators: Not reported.

Parasites: Not known.

IUCN conservation status: Data deficient.

3.2.6 GYMNURA CROOKI, Fowler, 1934

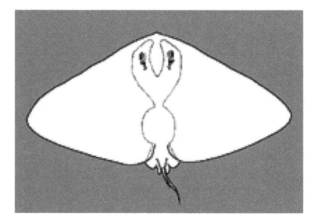

Common name: Not designated.

Geographical distribution: Subtropical; northwest Pacific: known from the type locality, Kowloon, China.

Habitat: Marine; demersal.

Distinctive features: Pectoral fins of this species extend in the form of wide wings that exceed the length of the disk. Snout is short and wide with a blunted tip. The spiracles are located after the eyes. On the ventral side of the disk, a fairly large curved mouth, nostrils, and five pairs of gill slits are seen. A leather flap lies between the nostrils. Teeth are small, narrow, and pointed. Pelvic fins are small and rounded.

Food habits: Not reported.

Reproduction: This species shows ovoviparity (aplacental viviparity), with embryos feeding initially on yolk. Additional nourishment for the embryos is from the mother by indirect absorption of uterine fluid (enriched with mucus, fat, or protein) through specialized structures.

Parasites: Not known.

IUCN conservation status: Not reported.

3.2.7   *GYMNURA HIRUNDO* (Lowe, 1843)

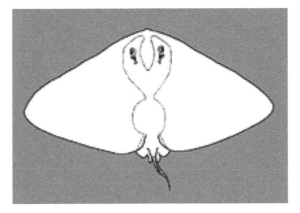

Common name: Madeira butterfly ray.

Geographical distribution: Subtropical; eastern central Atlantic: Madeira Islands. Southwest Atlantic: southern Brazil.

Habitat: Marine; demersal.

Distinctive features: Not much is known about the biology of this species. This species grows to a maximum size of 167 cm DW.

Food habits: Not known.

Reproduction: This species shows ovoviparity (aplacental viviparity), with embryos feeding initially on yolk. Additional nourishment for the embryos is from the mother by indirect absorption of uterine fluid (enriched with mucus, fat, or protein) through specialized structures.

Parasites: Not reported.

IUCN conservation status: Not evaluated.

3.2.8 *GYMNURA JAPONICA* (Temminck & Schlegel, 1850)

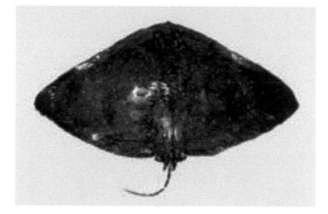

Common name: Japanese butterflyray.

Geographical distribution: Tropical; northwest Pacific: southern Japan to the China seas; Thailand and India.

Habitat: Marine; demersal; inhabits sandy or muddy bottom in shallow waters.

Distinctive features: Body of this species is flattened and is surrounded by an extremely broad disk formed by the pectoral fins. It has a very short, thread-like, tail. Average STL of males is 8 mm and 15 mm in females. In both the sexes, average number of total serrations is 21. It attains a maximum size of 145 cm DW (Schwartz, 2007).

Food habits: Feeds on benthic animals.

Reproduction: Exhibit ovoviparity (aplacental viviparity), with embryos feeding initially on yolk. Additional nourishment for the embryos is from the mother by indirect absorption of uterine fluid (enriched with mucus, fat, or protein) through specialized structures. Maturing males of this species measure 44 cm DW (Masuda et al., 1984).

Predators: Not known.

Parasites (Izawa, 2011)

Copepoda: *Eudactylina gymnuri*

IUCN conservation status: Data deficient.

3.2.9 *GYMNURA MARMORATA* (Cooper, 1864)

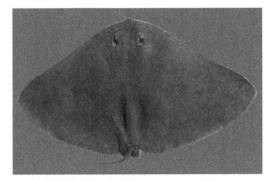

Common name: California butterfly ray.

Geographical distribution: Warm temperate to tropical; subtropical; eastern Pacific: California, USA to Peru.

Habitat: Marine; demersal; shallow bays and beaches.

Distinctive features: Front margin of the disk of this species is slightly concave and the rear margin is rounded. Snout is blunt except for a protruding tip. Short tail is approximately one-half the length of the disk and there is no caudal fin. Distance between the eyes is greater than the distance from the eyes to the tip of the snout. Eyes are located at 16% of the body length. This species has brown in color and at times, it is covered with small brown or black spots. Width of the disk is 1.5 times the length, reaching a maximum size of 1.5 m wide and 1 m long.

Food habits: Feeds on crustaceans and small fishes.

Reproduction: This species shows ovoviparity (aplacental viviparity), with embryos feeding initially on yolk. Additional nourishment for the embryos is from the mother by indirect absorption of uterine fluid (enriched with mucus, fat, or protein) through specialized structures. The reproductive cycle in a year shows a gestation period of 9–12 months and birth size is 23 cm DW.

Predators: Not known.

Parasites

Cestoda: *Acanthobothrium parviuncinatum*

Copepoda: *Trebius latifurcatus*

IUCN conservation status: Least concern.

3.2.10 *GYMNURA MICRURA* (Bloch & Schneider, 1801)

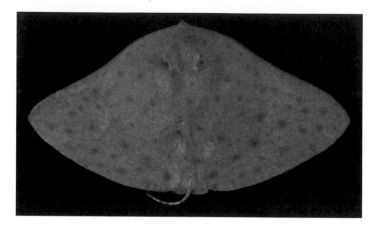

Common name: Smooth butterfly ray.

Geographical distribution: Western Atlantic: Maryland, USA to Brazil; Gulf of Mexico and northern South America to Brazil; eastern Atlantic: Senegal, Gambia, Sierra Leone, Cameroon and Democratic Republic of the Congo; Sri Lanka, India, Malaysia, Singapore, Viet Nam, Borneo, and Sumatra in Indonesia.

Habitat: Marine; brackish; demersal; deep-water; prefers neritic waters of the continental shelf and usually found on soft bottoms; brackish estuaries or hyper-saline lagoons.

Distinctive features: This species has a broad, diamond-shaped body with a very short tail lacking a dorsal spine. Snout is protruding. Front edges of disk are concave. Tail is with low dorsal and ventral fin folds and 3–4 dark crossbars. Upper surface is gray, brown, light green, or purple with round spots. Lower surface is white. This species grows to a maximum size of 137 cm DW.

Food habits: Feeds on fish and invertebrates, more specifically bony fish, benthic copepods, crabs, shrimps, prawns, and bivalves.

Reproduction: This species shows ovoviparity (aplacental viviparity), with embryos feeding initially on yolk. Additional nourishment for the embryos is from the mother by indirect absorption of uterine fluid (enriched with mucus, fat, or protein) through specialized structures (McEachran & Séret, 1990).

Predators: Potential predators include larger fishes and marine mammals (FLMNH Ichthyology Dept., http://www.flmnh.ufl.edu/index.php?cID=1885).

Parasites

Cestoda: *Acanthobothrium fogeli*
Acanthobothrium giganticum
Acanthobothrium micracantha
Acanthobothrium somnathii
Hexacanalis abruptus
Hexacanalis pteroplateae
Halysiorhynchus macrocephalus
Hexacanalis pteroplateae
Phyllobothrium bifidum
Pterobothrium lesteri
Pterobothrium lintoni
Pterobothrium southwelli
Pterobothrium sp.
Uncibilocularis somnathii
IUCN conservation status: Data deficient.

3.2.11   GYMNURA NATALENSIS (Gilchrist & Thompson, 1911)

Common name: Butterfly ray.

Geographical distribution: Subtropical; southeast Atlantic and western Indian Ocean: Namibia round the Cape to southern Mozambique.

Habitat: Marine; demersal; sandy beaches, muddy estuaries, and offshore banks.

Distinctive features: This species has a broad diamond-shaped pectoral disk which is almost twice as wide as it is long. Tail is shorter than body with black and white bands and a small sting. Tentacle is seen at rear edge of each spiracle. Body coloration is gray, green, or brown above, often with darker mottling and white below. It can change color of upper disk quickly to match substrate. It grows to a maximum size of 250 cm DW and weight is 82.6 kg (Compagno, 1986).

Food habits: Feeds on a variety of fishes, crabs and polychaete worms.

Reproduction: This species shows ovoviparity (aplacental viviparity), with embryos feeding initially on yolk. Additional nourishment for the embryos is from the mother by indirect absorption of uterine fluid (enriched with mucus, fat, or protein) through specialized structures.

Predators: Not known.

Parasites

Copepoda: *Schistobrachia jordaanae*

IUCN conservation status: Data deficient.

3.2.12   *GYMNURA POECILURA* (Shaw, 1804)

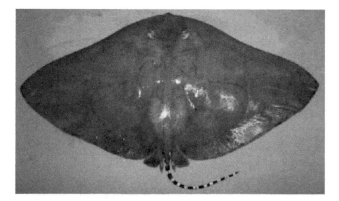

Common name: Long-tailed butterfly ray.

Geographical distribution: Tropical; Indo-Pacific: Red Sea to French Polynesia, north to Japan.

Habitat: Marine; demersal; sandy bottoms of shallow inshore waters and offshore banks.

Distinctive features: Pectoral fin disk of this species is rhomboid measuring around twice as wide as long. Leading margin of the disk is gently sinuous and trailing margin is convex. Snout is short and broad, with a tiny protruding tip. Medium-sized eyes have larger, smooth-rimmed spiracles behind. Large mouth forms a transverse curve and contains over 50 tooth rows in each jaw. Teeth are small, narrow, and pointed. There are five pairs of short gill slits on the underside of the disk. Pelvic fins are small and rounded. Thread-like tail lacks dorsal or caudal fins, though there are low ridges along its length above and below. Sometimes, there is a small stinging spine (very rarely two) on the upper surface of the tail near the base. Skin is devoid of dermal denticles. Coloration of this species is brown to greenish brown to gray above, with many small pale spots and sometimes with dark dots. Tail has 9–12 black bands alternating with white bands, which often have a small, dorsally positioned dark spot within. Belly is white, darkening at the edges of the fins. This species may attain a maximum size of 250 cm DW.

Food habits: Feeds on bony fishes (ponyfishes of the genus *Leiognathus*), molluscs, and crustaceans.

Reproduction: This species shows ovoviparity (aplacental viviparity), with embryos feeding initially on yolk. Additional nourishment for the embryos is from the mother by indirect absorption of uterine fluid (enriched with mucus, fat, or protein) through specialized structures. Reproductive activity proceeds throughout the year and gestation period is unknown, Litter size is seven. Newborns measure 20–26 cm across. Males and females mature sexually at roughly 45 and 41 cm across, respectively. Adult females have two functional ovaries and uteruses.

Predators: Not known.

Parasites

Cestoda: *Acanthobothrium micracantha*

Pterobothrium lesteri

Proemotobothrium southwelli

Nematoda: *Hysterothylacium poecilurai*

IUCN conservation status: Near threatened.

3.2.13 GYMNURA TENTACULATA (Müller & Henle, 1841)

Common name: Tentacled butterfly ray.

Geographical distribution: Tropical; western central Pacific: Papua New Guinea.

Habitat: Marine; demersal.

Distinctive features: Outer anterior margin of pectorals of this species is continuous alongside of head. Disk is extremely broad. Dorsal fin and tail spines are present. Tail is short. It reaches a maximum size of 23.5 cm total length.

Food habits: Very little is known of the habitat and ecology of this species (Jacobsen, 2009).

Reproduction: This species shows ovoviparity (aplacental viviparity), with embryos feeding initially on yolk. Additional nourishment for the embryos is from the mother by indirect absorption of uterine fluid (enriched with mucus, fat, or protein) through specialized structures.

Predators: Not known.

Parasites: Not reported.

IUCN conservation status: Data deficient.

3.2.14   *GYMNURA ZONURA* (Bleeker, 1852) = *Aetoplatea zonura*

Common name: Zonetail butterfly ray.

Geographical distribution: Tropical; eastern Indian Ocean and western central Pacific: India to Indonesia, including Singapore and Thailand.

Habitat: Marine; reef-associated; demersal inshore; shallow depths.

Distinctive features: The average STL of this species is 15 mm in males and 17 mm in females. Average total spine serrations are 50 for males and 28 for females. A small but noticeable dorsal fin is seen at base of tail. Dorsal surface of disk is brown with numerous dark spots or lines interspersed with large round yellowish spots or irregular blotches, sometimes forming lines or ocelli. Maximum total length is 65 cm and maximum width is 85 cm (Schwartz, 2007).

Food habits: Feeds on benthic crustaceans and teleost fishes.

Reproduction: This species shows ovoviparity (aplacental viviparity), with embryos feeding initially on yolk. Additional nourishment for the embryos is from the mother by indirect absorption of uterine fluid (enriched with mucus, fat, or protein) through specialized structures. Males mature at about 48 cm DW. Female gives birth to four pups (White, 2006b).

Predators: Not known.

Parasites

Cestoda: *Hexacanalis folifer*

IUCN conservation status: Vulnerable.

3.3   EAGLE RAYS/MANTA RAYS (MYLIOBATIDAE)

3.3.1   *AETOBATUS FLAGELLUM* (Bloch & Schneider, 1801)

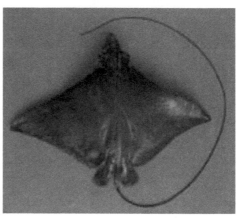

Phylum: Chordata   Subphylum: Vertebrata
Class: Chondrichthyes   Subclass: Elasmobranchii
Order: Myliobtiformes   Family: Myliobatidae

Common name: Longheaded eagle ray.

Geographical distribution: Tropical; Indo-West Pacific: Red Sea, India, East Indies, and southern China, possibly in the eastern Atlantic.

Habitat: Marine (inshore waters); brackishwater; benthopelagic; amphidromous.

Distinctive features: Tail and stinging spines of this species are long. Average STL of females is 42 mm with total spine serrations as 115. Head is long. Rostral lobe may be long to very long (longest in adult males) with a narrowly pointed apex. Teeth plates are in a single row. Teeth of lower jaw are chevron shaped. Dorsal surface of this species is uniformly brownish, without pale spots. This species grows to a maximum total length of 126 cm, maximum length of 90 cm DW, and weight of 13.9 kg (Schwartz, 2007; White, 2006a; White & Moore, 2013).

Food habits: Diet consists of a wide variety of benthic species including polychaetes, bivalve and gastropod molluscs, cephalopods, crustaceans, and teleost fishes.

Reproduction: This species exhibits ovoviparity (aplacental viviparity), with embryos feeding initially on yolk. Additional nourishment for the embryos is from the mother by indirect absorption of uterine fluid (enriched with mucus, fat, or protein) through specialized structures. Males mature by about 50 cm DW and females by about 75 cm DW. Average annual fecundity or litter size is up to 4 (McEachran & Séret, 1990).

Predators: Not reported.

Parasite

Cestoda: *Polypocephalus bombayensis* (Koch et al., 2012).

IUCN conservation status: Endangered.

3.3.2 *AETOBATUS NARINARI* (Euphrasen, 1790) = *Aetobatus laticeps*

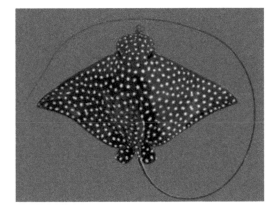

Common name: Spotted eagle ray.

Geographical distribution: Subtropical; western Atlantic: North Carolina and Florida, USA and Bermuda to southern Brazil; Gulf of Mexico and Caribbean; eastern Atlantic: Mauritania to Angola; Indo-West Pacific: Red Sea and South Africa to Hawaii, north to Japan, south to Australia; Eastern Pacific: Gulf of California to Puerto Pizarro, Peru, and the Galapagos Islands.

Habitat: Marine; brackishwater; shallow inshore waters such as bays and estuaries; reef-associated; amphidromous.

Distinctive features: This species has a long snout which is flat and rounded like a duck's bill. Head is thick and its pectoral disk is sharply curved. Caudal fin is absent. Jaws are usually with single row of flat, chevron-shaped teeth. Each tooth is a crescent-shaped plate joined into a band. Numerous white spots are seen on the black or bluish disk. Belly is white. Tail is long and whip like with a long spine near the base. Average total spine serrations are 102 for males and 113 for females. Disk is without spines. It grows to a maximum size of 330 cm DW and weight of 230 kg (Schwartz, 2007).

Food habits: Feeds mainly on bivalves but also eats shrimps, crabs, octopus, worms, whelks, and small fish.

Reproduction: This species exhibits ovoviparity (aplacental viviparity), with embryos feeding initially on yolk. Additional nourishment for the embryos is from the mother by indirect absorption of uterine fluid (enriched with mucus, fat, or protein) through specialized structures. It bears up to four young.

Preators: Not known.

Parasites (Koch et al., 2012)

Monogenea: *Clemacotyle australis*

Decacotyle octona

Decacotyle elpora

Dendromonocotyle torosa

Cestoda: *Acanthobothrium aetiobatis*

Acanthobothrium arlenae

Acanthobothrium colombianum

Acanthobothrium dysbiotos

Acanthobothrium nicoyaense

Acanthobothrium monski

Acanthobothrium tortum

Adelobothrium aetiobatidis

Cephalobothrium aetobatidis

Didymorhynchus southwelli

Disculiceps sp.

Dollfusiella aetobati

Echinobothrium boisii

Eutetrarhynchus aetobati

Hornellobothrium cobraformis

Hornellobothrium extensivum

Kotorella pronosoma
Kystocephalus translucens
Myzophyllobothrium rubrum
Oncomegas australiensi
Oncomegas sp.
Parachristianella baverstocki
Proemotobothrium linstowi
Shirleyrhynchus aetobatidis
Staurobothrium aetobatidis
Trimacracanthus aetobatidis
Trygonicola macroporus
Tylocephalum aurangabadensis
Tylocephalum girindrai
Nematoda: *Echinocephalus sinensis*
Echinocephalus overstreeti
Terranova scoliodontis
Copepoda: *Eudactylina hornbosteli*

IUCN conservation status: Near threatened.

3.3.3 AETOBATUS NARUTOBIEI, White et al., 2013

Common name: Naru eagle ray.

Geographical distribution: Subtropical; north-west Pacific: eastern Vietnam, Hong Kong, China, Korea, and southern Japan.

Habitat: Marine (shallower bays); benthopelagic.

Distinctive features: Head of this species is long. Rostral lobe is long to very long (longest in adult males) and is narrow, tapering evenly to tip. There is a single row of teeth plates and those in lower jaw are chevron shaped. It has uniformly greenish gray to brownish dorsal surfaces, without pale spots or ocelli. This species grows to a maximum total length of 150 cm, maximum DW of 100 cm, and weight of 14.4 kg (White et al., 2013).

Food habits: Feeds primarily on shellfish.

Reproduction: This species exhibits ovoviparity (aplacental viviparity), with embryos feeding initially on yolk. Additional nourishment for the embryos is from the mother by indirect absorption of uterine fluid (enriched with mucus, fat, or protein) through specialized structures.

Predators: Not known.

Parasites: Not reported.

IUCN conservation status: A new species—status not known.

3.3.4   *AETOBATUS OCELLATUS* (Kuhl, 1823)

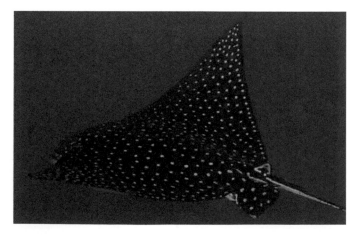

Common name: Ocellated eagle ray, whitespotted eagle ray.

Geographical distribution: Tropical and subtropical zones; Indo-West Pacific.

Habitat: Marine; benthopelagic; coastal waters and estuarine habitats.

Distinctive features: This species has an angular shaped disk with white spots on the upper surface. Snout can vary in shape from spade-like in juveniles pointed in large individuals. Tail is whip like with 2–6 spines. It grows to a maximum size of 153 cm DW and weight of 200 kg.

Food habits: Feeds on gastropod, bivalve molluscs, crustraceans, worms, octopuses, and fishes.

Reproduction: It exhibits ovoviparity (aplacental viviparity), with embryos feeding initially on yolk. Additional nourishment for the embryos is from the mother by indirect absorption of uterine fluid (enriched with mucus, fat, or protein) through specialized structures.

Predators: Not known.

Parasites (White et al., 2010)

Monogenea: *Merizocotyle pseudodasybatis*

Cestoda: *Aetobatus ocellatus*

Echinobothrium boisii

Adelobothrium aetiobatidis

Calycobothrium typicum

Cephalobothrium aetobatidis

Hornellobothrium cobraformis

Hornellobothrium extensivum

Hornellobothrium sp.

Hornellobothrium sp.

Kystocephalus translucens

Staurobothrium aetobatidis

Tenia narinari

Tylocephalum aetiobatidis

Tylocephalum aurangabadensis

Tylocephalum girindrai

Tylocephalum yorkei

Acanthobothrium aetiobatis

Acanthobothrium arlenae

Acanthobothrium dysbiotos

Myzocephalus narinari

Myzophyllobothrium rubrum

Didymorhynchus southwelli

Dollfusiella aetobati

Kotorella pronosoma

Oncomegas aetobatidis

Oncomegas australiensis

Parachristianella baverstocki

Proemotobothrium linstowi

Shirleyrhynchus aetobatidis

Trygonicola macroporus

IUCN conservation status: Not evaluated.

3.3.5   AETOMYLAEUS MACULATUS (Gray, 1834)

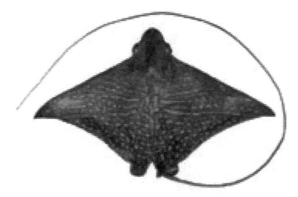

Phylum: Chordata   Subphylum: Vertebrata
Class: Chondrichthyes   Subclass: Elasmobranchii
Order: Myliobtiformes   Family: Myliobatidae

Common name: Mottled eagle ray.

Geographical distribution: Tropical; Indo-West Pacific: India to China and Indonesia.

Habitat: Marine; brackish; reef-associated; inshore waters, mangrove creeks, and protected sandy channels.

Distinctive features: This species has an exceptionally long spineless tail which is over six times longer than the body. Disk of this species is marked with brown-edged stripes. Dorsal fin originates behind ends of pelvic bases. It grows to a maximum size of 200 cm DW.

Food habits: Preys on a wider array of food items such as bivalves, shrimps, crabs, octopus, worms, whelks, and small fish.

Reproduction: It exhibits ovoviparity (aplacental viviparity), with embryos feeding initially on yolk. Additional nourishment of the embryos is from the mother by indirect absorption of uterine fluid (enriched with mucus, fat, or protein) through specialized structures. It produces an average of four offspring per year.

Predators: Not known.

Parasites (Chisholm & Whittington, 2004; Koch et al., 2012)

Monogenea: *Empruthotrema dasyatidis*

Myliocotyle borneoensis

Cestoda: *Elicilacunosus sarawakensis*

Diagonobothrium assymetrum

Elicilacunosus sarawakensis

Tylocephalum dierama

Discobothrium redacta

Dollfusiella spp.

Discobothrium quadrisurculi

Acanthobothrium myliomaculata

Anthobothrium panjadi

Rhoptrobothrium myliobatidis

IUCN conservation status: Endangered.

3.3.6 *AETOMYLAEUS MILVUS* (Müller & Henle, 1841)

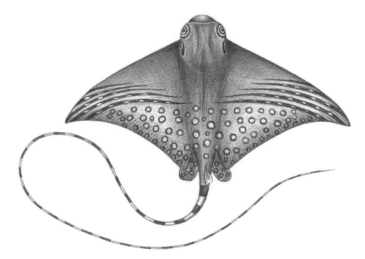

Common name: Mottled eagle ray.

Geographical distribution: Tropical; Indo-West Pacific: Red Sea to China, south to Indonesia.

Habitat: Marine; benthopelagic; inshore waters with sandy substrates.

Distinctive features: Disk of this species has a concave trailing edge, before a sharp angle. Head is wide and flat. Anterior nasal flaps are large. Mouth is horizontal and flat. Pelvic fins are narrow, about 2.5 times the width of the front corner. Tail is slender, about three times the body length. Skin fold is absent. Tail spines are absent. It has a maximum DW of 37 cm.

Food habits: It feeds on benthic molluscs, crustaceans, and fish.

Reproduction: It exhibits ovoviparity (aplacental viviparity), with embryos feeding initially on yolk. Additional nourishment for the embryos is from the mother by indirect absorption of uterine fluid (enriched with mucus, fat, or protein) through specialized structures. Male matures at 71 cm DW. It has low fecundity and bears litters of up to four offspring.

Predators: Not known.

Parasites: Not reported.

IUCN conservation status: Endangered.

3.3.7   *AETOMYLAEUS NICHOFII* (Bloch & Schneider, 1801)

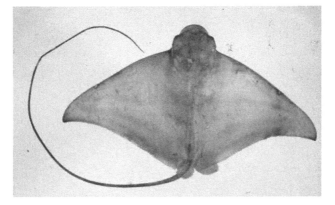

Common name: Banded eagle ray.

Geographical distribution: Tropical; Indo-West Pacific: Persian Gulf to the Philippines, north to Korea and southern Japan, south to northern Australia, possibly in the Red Sea and eastern and southern Africa.

Habitat: Marine; brackish; demersal; amphidromous; inshore and offshore, from the intertidal to at least 70 m depth.

Distinctive features: Disk of this species is very broad and its length is 1.67 in its width. Snout is short, and it forms pointed lobe at low level in front. Eyes are not visible from above. Spiracles are large, deep, and open laterally. Teeth are with broad median row and three narrow lateral rows are seen on each side. Gill openings are small and subequal. One dorsal fin is small. Ventral fins are about twice as long as wide. Pectorals are subfalciform. Tail is long, slender, and whip like. Caudal spine is absent. Coloration is black brown with five or six lighter, bluish, narrow bands. Belly is whitish. It grows to a maximum size of 65.0 cm DW (Last & Stevens, 1994).

Food habits: Feeds on worms, crustaceans, snails, and bony fishes.

Reproduction: It exhibits ovoviparity (aplacental viviparity), with embryos feeding initially on yolk. Additional nourishment for the embryos is from the mother by indirect absorption of uterine fluid (enriched with mucus, fat, or protein) through specialized structures. It gives birth to four pups. Size at birth is 17 cm DW.

Predators: Not known.

Parasites (Chisholm & Whittington, 2004; Koch et al., 2012)

Monogenea: *Empruthotrema dasyatidis*

Myliocotyle multicrista

Empruthotrema dasyatidis

Cestoda: *Elicilacunosus dharmadii*

Elicilacunosus fahmii

Elicilacunosus dharmadii

Elicilacunosus fahmii

Acanthobothrium hanumantharaoi

Acanthobothrium rhynchobatidis

Myliobatibothrium alii

Myliobatibothrium singhi

Rhoptrobothrium chongi

Rhoptrobothrium gambangi

Rhoptrobothrium limae

Acanthobothrium rhynchobatidis

Prochristianella spp.

Dollfusiella spp.

IUCN conservation status: Vulnerable.

3.3.8   *AETOMYLAEUS VESPERTILIO* (Bleeker, 1852)

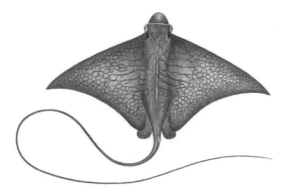

Common name: Ornate eagle ray, reticulate eagle ray.

Geographical distribution: Tropical; Indo-West Pacific: scattered localities, from southern Mozambique to the western Pacific.

Habitat: Marine; benthopelagic; muddy bays and banks and on coral reefs.

Distinctive features: This species has a clearly distinct pattern of reticulate dark lines and rings on its back. Rear margin of disk and pelvic fins are with white spots. Disk is as wide as long. Anterior margins are slightly convex and posterior margins are concave. Snout is long and produced. Head is thick and prominent. Eyes are prominent and are laterally placed. Dorsal fin is small and has its origin above the ends of the bases of the ventrals. Ventrals are narrow and their hind margin is rounded. Tail is long, without a caudal spine. Body is with small spines. It grows to a maximum size of 240 cm DW and body length of 4 m. It is deemed harmless to humans (Benjamin et al., 2012; Last & Stevens, 1994).

Food habits: Feeds on worms, crustaceans, snails, and bony fishes.

Reproduction: It exhibits ovoviparity (aplacental viviparity), with embryos feeding initially on yolk. Additional nourishment for the embryos is from the mother by indirect absorption of uterine fluid (enriched with mucus, fat, or protein) through specialized structures.

Predators: Not known.

Parasites (Koch et al., 2012)

Monogenea: *Malalophus jensenae*

Cestoda: *Collicocephalus baggioi*

Rexapex nanus

Aberrapex weipaensis

Aberrapex weipaensis

Collicocephalus baggio

Rexapex nanus

Copepoda: *Janinecaira darkthread*

IUCN conservation status: Endangered.

3.3.9 *MYLIOBATIS AQUILA* (Linnaeus, 1758)

Phylum: Chordata   Subphylum: Vertebrata
Class: Chondrichthyes   Subclass: Elasmobranchii
Order: Myliobtiformes   Family: Myliobatidae

Common name: Common eagle ray.

Geographical distribution: Subtropical; eastern Atlantic: Madeira, Morocco, Canary Islands north to the western coasts of Ireland and British Isles; south-western North Sea; South Africa; Mediterranean Sea.

Habitat: Marine; benthopelagic; shallow lagoons, bays and estuaries; offshore waters.

Distinctive features: Disk of this species is much wider with broad, angular corners. Head is moderately short and it has a bluntly rounded snout. Dorsal fin originates behind the tips of the pelvic fins. Pectoral fin rays continue forward under the eyes to form a subrostral lobe under the snout. Lobe is short and obtuse. Tail is slender and whip like with one or rarely two spines on the top of the tail close to the body. Caudal fin is absent. Dorsal surface is dusky bronze or blackish with olive to violet shading. Ventral surface is white with a brownish margin. Tail is blackish. Maximum reported size of this species is a DW of 183 cm and maximum reported weight is 14.5 kg.

Food habits: Feeds on benthic crustaceans, molluscs, and fish.

Reproduction: This species shows ovoviparity (aplacental viviparity), with embryos feeding initially on yolk. Additional nourishment for the embryos is from the mother by indirect absorption of uterine fluid (enriched with mucus, fat, or protein) through specialized structures. Males reach sexual maturity at 40–50 cm total length while females mature at 60–70 cm total length. Gestation period is about 8 months and 3–7 young are produced. Newly born rays closely resemble adults in morphology (McEachran & Séret, 1990).

Predators: Not known.

Parasites (Koch et al., 2012)

Monogenea: *Monocotyle myliobatis*

Cestoda: *Acanthobothrium batailloni*

Caulobothrium longicolle

Christianella trygonbrucco

Dollfusiella spinifer

Echinobothrium mathiasi

Echeneibothrium myliobatis

Echinobothrium typus

Parachristianella monomegacantha

Parachristianella trygonis

Prochristianella glaber

Rhinebothrium setiensis

Rhodobothrium lubeti

Tetrarhynchobothrium setiense

Tetrarhynchobothrium striatum

Tetrarhynchobothrium tenuicolle

IUCN conservation status: Data deficient.

3.3.10 *MYLIOBATIS AUSTRALIS,* Macleay, 1881

Common name: Australian bull ray

Geographical distribution: Subtropical; eastern Indian Ocean: southern Australia, from Western Australia to Queensland; New Zealand.

Habitat: Marine; reef-associated.

Distinctive features: This species has a disk which is wider than it is long. Snout is blunt with a skirt-shaped internasal flap as well as a single fleshy lobe. Medium-sized eyes are located on the sides of the head and the spiracles are large. Pectoral fins which make up the wings of this species originate below the eyes. Margins of the pectoral fins are deeply concave and the tips are highly angular. There is a small dorsal fin. Tail is elongate and whip like and has a venomous stinging spine located just behind the dorsal fin. Dorsal surface is brownish-gray to olive-green or yellowish with bluish spots and crescent-shaped bars. It is paler on the ventral surface sometimes with a grayish margin on the disk. Maximum DW and weight of this species are 1.2 m and 56.5 kg, respectively (FLMNH Ichthyology Dept., http://www.flmnh.ufl.edu/index.php?cID=2038).

Food habits: Feeds mainly on crabs and shellfish.

Reproduction: This species shows ovoviparity (aplacental viviparity), with embryos feeding initially on yolk. Additional nourishment for the embryos is from the mother by indirect absorption of uterine fluid (enriched with mucus, fat, or protein) through specialized structures. Males reach sexual maturity at a DW of 0.65 m with females becoming mature at a DW of 0.80 m. Each female gives birth to litters of 2–15 (average 6) young.

Although reproductive periodicity is not known, it is believed to be 1–3 years in length.

Predators: Potential predators of the Australian bull ray include larger fish such as sharks as well as marine mammals.

Parasites (Koch et al., 2012)

Myxozoa: *Chloromyxum myliobati*

Monogenea: *Heliocotyle ewingi*

Cestoda: *Acanthobothrium martini*

Acanthobothrium pichelinae

Eutetrarhynchus bareldsi

Eutetrarhynchus martini

Eutetrarhynchus ocallaghani

Dollfusiella bareldsi

Dollfusiella martini

Dollfusiella ocallaghani

Parachristianella monomegacantha

Tetrarhynchobothrium australe

Trimacracanthus aetobatidis

IUCN conservation status: Least concern.

3.3.11 *MYLIOBATIS CALIFORNICUS,* Gill, 1865

Common name: Bat eagle ray.

Geographical distribution: Subtropical; Eastern Pacific: Oregon, USA to Gulf of California and Galapagos Islands.

Habitat: Marine; demersal; sandy and muddy bays and sloughs; rocky bottom and in kelp beds.

Distinctive features: This species has a flat body with a distinct protruding head, large eyes, and smooth skin. Whip-like tail is as long or longer than the body width with a dorsal fin at the base and armed with a barb-like spine located just behind the body. Its long pectoral fins resemble bat wings. It lacks arm-like cephalic fins. It has smooth skin which is dark brown or black in color, changing to white on the underside. Maximum DW of this species is 180 cm and maximum weight is 91 kg.

Food habits: Feeds on bivalves, snails, polychaetes, shrimps, and crabs.

Reproduction: This species shows ovoviparity (aplacental viviparity), with embryos feeding initially on yolk. Additional nourishment for the embryos is from the mother by indirect absorption of uterine fluid (enriched with mucus, fat, or protein) through specialized structures. Males reach maturity at DWs of 67–68 cm and weights of 5 kg while females mature at larger widths than males and weights of 23 kg. Following a gestation period of 8–12 months, female enters shallow waters to give birth. Litter size is ranging from 2 to 12 pups. At birth, pups are 30–36 cm in width and each weighs about 1 kg. At birth, young bat rays have a DW of about 3 m.

Predators: Sharks, including the white shark (*Carcharodon carcharias*) and seven-gill shark (*Notorynchus cepedianus*), feed on bat rays. Sea lions consume immature bat rays.

Parasites (Koch et al., 2012)

Monogenea: *Dendromonocotyle californica*

Cestoda: *Echinobothrium mexicanum*

Echinobothrium fautleyae

Aberrapex senticosus

Discobothrium myliobatidis

Echeneibothrium opisthorchis

Rhodobothrium brachyascum

Acanthobothrium holorhini

Acanthobothrium maculatum

Acanthobothrium microcephalum

Acanthobothrium unilateralis

Acanthobothrium myliobatidis

Aberrapex senticosus

Caulobothrium tetrascaphium

Mecistobothrium myliobati

Dollfusiella sp.

Parachristianella sp.

Prochristianella sp.

Trematoda: *Probolitrema richiardii*

Copepoda: *Trebius latifurcatus*

IUCN conservation status: Least concern.

3.3.12 *MYLIOBATIS CHILENSIS,* Philippi, 1892

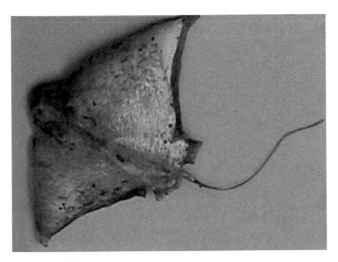

Common name: Chilean eagle ray.

Geographical distribution: Subtropical; southeast Pacific: Peru to Golfo de Arauco, Chile.

Habitat: Marine; benthopelagic; inshore on the continental shelf.

Distinctive features: Nothing is known of its biology. This species grows to a maximum size of 200 cm DW (Lamilla, 2006a).

Food habits: Benthic feeder.

Reproduction: This species shows ovoviparity (aplacental viviparity), with embryos feeding initially on yolk. Additional nourishment for the embryos is from the mother by indirect absorption of uterine fluid (enriched with mucus, fat, or protein) through specialized structures.

Predators: Not known.

Parasites (Koch et al., 2012)

Cestoda: *Rhodobothrium mesodesmatum*

Caulobothrium myliobatidis

Acanthobothrium batailloni

Acanthobothrium coquimbensis

IUCN conservation status: Data deficient.

3.3.13   MYLIOBATIS FREMINVILLII, Lesueur, 1824

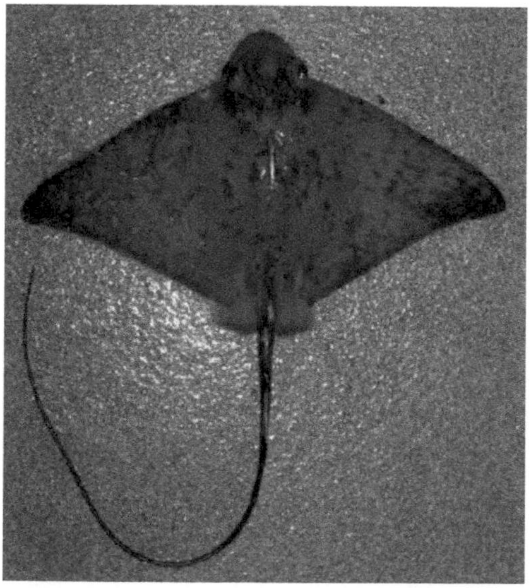

Common name: Bullnose eagle ray.

Geographical distribution: Subtropical; western Atlantic: Cape Cod to south-eastern Florida; southern Brazil to Argentina; Gulf of Mexico, Florida, and Caribbean islands, and northern South America.

Habitat: Marine; brackish; benthopelagic; coastal waters, mainly in shallow estuaries.

Distinctive features: Disk of this species is broad with long and sharply pointed wings. Snout projects distinctly from the body. Eyes and spiracles are located on the sides of the head. There are few middorsal spines on the disk. Corners of pectoral fins are acute angled. Dorsal fin originates close to the level of the pelvic fin rear margins. Whip-like tail is quite long and distinctly marked off from the body. There is no caudal fin. Dorsal surface is gray, chocolate, or brown while the ventral surface is pale or white. There may be whitish or yellowish small spots across the dorsal surface. Dorsal fin is rarely pale in color and the posterior portion of the tail is dark brown or black. Teeth are green. Maximum size of this species is 100 cm DW (FLMNH Ichthyology Dept., http://www.flmnh.ufl.edu/index.php?cID=2041).

Food habits: Feeds on benthic invertebrates including crustaceans and molluscs.

Reproduction: This species shows ovoviparity (aplacental viviparity), with embryos feeding initially on yolk. Additional nourishment for the embryos is from the mother by indirect absorption of uterine fluid (enriched with mucus, fat, or protein) through specialized structures. Males mature sexually at 60–70 cm DW. A female produces four to eight embryos during a reproductive season. Each newborn has a DW of 25 cm and closely resembles the adults in morphology.

Predators: Potential predators of this ray include marine mammals and large fish such as sharks.

Parasites (Koch et al., 2012)

Cestoda: *Nybelinia* sp.

Tetrarhynchobothrium unionifactor

Caulobothrium longicolle

IUCN conservation status: Data deficient.

3.3.14   *MYLIOBATIS GOODEI,* Garman, 1885

Common name: Southern eagle ray

Geographical distribution: Worldwide in tropical waters. Western Atlantic: South Carolina, USA to Argentina.

Habitat: Marine; benthopelagic.

Distinctive features: In this species, there is a small dorsal fin which is set farther back on tail, well beyond the pelvic fins. Disk is broader, with more rounded corners of wings. Snout is less projecting. There are no spines on disk. Broader separation is seen between the inner ends of gill openings. Body coloration is chocolate or grayish brown above and brownish white below. This species grows to a maximum total length of 125 cm (Robins & Ray, 1986).

Food habits: Its food consists of crustaceans and bivalves.

Reproduction: This species shows ovoviparity (aplacental viviparity), with embryos feeding initially on yolk. Additional nourishment for the embryos is from the mother by indirect absorption of uterine fluid (enriched with mucus, fat, or protein) through specialized structures.

Predators: Not reported.

Parasites (Koch et al., 2012)

Cestoda: *Echinobothrium megacanthum*

Aberrapex arrhynchum

Caulobothrium ostrowskiae

Caulobothrium uruguayensis

Phyllobothrium myliobatidis

Phyllobothrium sp.

Acanthobothrium sp.

IUCN conservation status: Data deficient.

3.3.15   *MYLIOBATIS HAMLYNI,* Ogilby, 1911

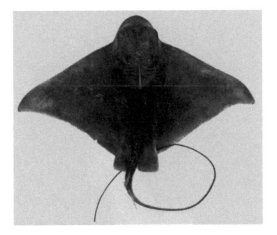

Common name: Purple eagle ray.

Geographical distribution: Eastern Indian Ocean and western central Pacific: Forestier Island, Western Australia and Cape Moreton, Queensland in Australia.

Habitat: Marine; benthopelagic; deep waters; rare offshore species.

Distinctive features: It is a small sized species. Disk is very broad which is plain on the dorsal surface, purplish to greenish gray centrally, and more olive brown laterally. Tail is long (1.5 times the width of disk) with a prominent stinging spine. This species has a maximum size of 59 cm DW.

Food habits: Nothing known of its biology (Kyne & Last, 2006).

Reproduction: This species shows ovoviparity (aplacental viviparity), with embryos feeding initially on yolk. Additional nourishment for the embryos is from the mother by indirect absorption of uterine fluid (enriched with mucus, fat, or protein) through specialized structures. It has low fecundity and each female gives birth up to four pups.

Predators: Not reported.

Parasites (Koch et al., 2012)

Cestoda: *Paroncomegas myliobati*

IUCN conservation status: Endangered.

3.3.16 *MYLIOBATIS LONGIROSTRIS,* Applegate & Fitch, 1964

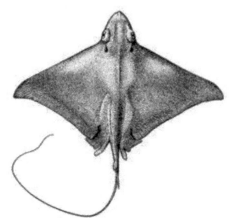

Common name: Longnose eagle ray, snouted eagle ray.

Geographical distribution: Subtropical; eastern Pacific: Gulf of California to Sechura, Peru.

Habitat: Marine; benthopelagic; sandy bottoms.

Distinctive Features: This species has a flat body with a large bulbous elevated blunt head and a long pointed snout. Disk is wider than it is long. It is dark reddish brown in coloration on top and the undersides are dusky. Head and eyes are behind the origin of the pectoral fins which are long and form equivalent triangles with pointed tips. There is a single dorsal fin. Long whip-like slender tail is equal in length to the length of the disk and it has a long spine at the base. It reaches a maximum size of 95 cm DW and weight of 20 kg (Mexico Fish, Flora, & Fauna, http://www.mexican-fish.com/longnose-eagle-ray/; Smith & Bizzarro, 2006).

Food habits: Feeds on small crustaceans, small fishes, mussels, and worms.

Reproduction: This species shows ovoviparity (aplacental viviparity), with embryos feeding initially on yolk. Additional nourishment for the embryos is from the mother by indirect absorption of uterine fluid (enriched with mucus, fat, or protein) through specialized structures. Size at maturity is 74 cm DW and 54 cm DW for males and females, respectively. Average annual fecundity or litter size of this species is unknown.

Predators: Not reported.

Parasites (Koch et al., 2012)

Cestoda: *Echinobothrium mexicanum*

Pseudochristianella nudiscula

Trematoda: *Probolitrema richiardii*

IUCN conservation status: Near threatened.

3.3.17   *MYLIOBATIS PERUVIANUS,* Garman, 1913

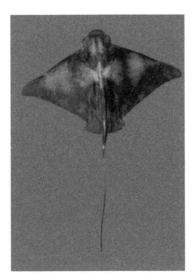

Common name: Peruvian eagle ray.

Geographical distribution: Subtropical; southeast Pacific: Peru and Chile.

Habitat: Marine; benthopelagic; soft bottoms.

Distinctive features: Disk of this species is rhomboid and is wider than long. Pectoral fins are distinctly pointed, concave at rear and, continue onto rostrum. Head and snout are blunt, rounded, raised, broad and are projecting well before disk. Eyes and spiracles are on side of head. Tail is slender about as long as disk. There is no tail fin. Tail has one large spine at its base after the small dorsal fin. Pointed pectoral lips are seen. Body coloration is dark brown above and, belly is white grading to brown on pectorals and pelvics. This species has a maximum total length and DW of 100 and 67 cm, respectively.

Food habits: Essentially nothing is known of its biology (Lamilla, 2006b).

Reproduction: This species shows ovoviparity (aplacental viviparity), with embryos feeding initially on yolk. Additional nourishment for the embryos is from the mother by indirect absorption of uterine fluid (enriched with mucus, fat, or protein) through specialized structures. It has low fecundity as with other myliobatids and litter size is up to four.

Predators: Not reported.

Parasites: Not known.

IUCN conservation status: Data deficient.

3.3.18   MYLIOBATIS RIDENS, Ruocco et al., 2012

Common name: Shortnose eagle ray.

Geographical distribution: Subtropical; southwest Atlantic: Brazil to Argentina.

Habitat: Marine; brackish; benthopelagic.

Distinctive features: This species has a rhombic disk which is wider than long. Head is clearly protruding from disk. Snout is short. Eyes are relatively small, lateral, greatly elevated, and strongly protruding. Spiracles are tear shaped, lateral and greatly enlarged. Mouth is relatively wide. Teeth are flattened and arranged like pavement stones. Normally, there are seven series of teeth in each jaw and teeth of middle series are much larger than those located toward corners of mouth. Dorsal fin is small. Pectoral fins are with weakly convex anterior margins; posterior margins are moderately concave, and outer corners are moderately rounded. Tail is elongate and whip like. It is markedly differentiated from body; much longer than disk, and without longitudinal folds or ridges. One or two serrated tail spines are seen close behind dorsal fin. Tail spines are serrated with lateral teeth and barbed tip. Skin is smooth and lacks denticles or thorns dorsally and ventrally. Claspers of adult males are moderately stout and cylindrical. Dorsal surface is dark brown to dark olive-green or orange-brown. Ventral surface is whitish with darker-orange or black margins of pectoral fins. Lower surface of pelvic fins, claspers, and tail is white. Tail is paler near base and is becoming darker to nearly black toward tip. Maximum size is 700 and 630 mm DW for females and males, respectively (Ruocco et al., 2012).

Food habits: Feeds on small crustaceans, small fishes, mussels, and worms.

Reproduction: This species shows ovoviparity (aplacental viviparity), with embryos feeding initially on yolk. Additional nourishment for the embryos is from the mother by indirect absorption of uterine fluid (enriched with mucus, fat, or protein) through specialized structures. Both sexes mature at approximately 500–600 mm DW.

Predators: Not reported.

Parasites: Not reported.

IUCN conservation status: Not evaluated.

3.3.19 MYLIOBATIS TENUICAUDATUS, Hector, 1877

Common name: Southern eagle ray, New Zealand eagle ray.

Geographical distribution: Temperate; southwest Pacific: Norfolk Island and New Zealand, including the Kermadec Islands.

Habitat: Marine; benthopelagic; bays, estuaries, and near rocky reefs.

Distinctive features: This species is olive-green, yellow, or dark brown dorsally with pale blue or gray markings. Belly is white. There is a spine on the short tail. There is no caudal fin. It has a maximum size of 150 cm DW.

Food habits: Feeds on clams, oysters, worms, and crabs.

Reproduction: This species shows ovoviparity (aplacental viviparity), with embryos feeding initially on yolk. Additional nourishment for the embryos is from the mother by indirect absorption of uterine fluid (enriched with mucus, fat, or protein) through specialized structures. Size at maturity is unknown. Litter size and size at birth are poorly known. Female gives birth to a litter of 20 pups after 6 months in captivity. Disk length at birth is 8 cm (Duffy, 2003).

Predators: Killer whales and white shark (*Carcharodon carcharias*).

Parasites (Koch et al., 2012)

Cestoda: *Trimacracanthus aetobatidis*

IUCN conservation status: Least concern.

3.3.20 *MYLIOBATIS TOBIJEI,* Bleeker, 1854

Common name: Japanese eagle ray.

Geographical distribution: Tropical; northwest Pacific: Japan, Korea, Okinawa Trough, China, and South China Sea; Philippines; Indonesia.

Habitat: Marine; demersal; inshore and offshore, intertidal habitats.

Distinctive features: In this species, tip of snout forms an obtuse angle. Body panel is mottle which is not however noticeable in the uniformly dark brown. Tail is long with a thorn. Average STL of males is 42 mm and 28 mm in females. Total number of serrations in these sexes is 60 and 42, respectively. Dorsal fin is in the rear position than the belly fin. This species attains 114 cm DW and 150 cm total length (Schwartz, 2007; http://zukan.com/fish/internal842).

Food habits: Carnivore feeding on benthic shrimp, crab, fish, and clams.

Reproduction: This species shows ovoviparity (aplacental viviparity), with embryos feeding initially on yolk. Additional nourishment for the embryos is from the mother by indirect absorption of uterine fluid (enriched with mucus, fat, or protein) through specialized structures. Males mature by 65 cm DW and females produce eight young per litter (Jeong et al., 2009).

Predators: Not reported.

Parasite (Koch et al., 2012)

Cestoda: *Caulobothrium tobejei*

IUCN conservation status: Data deficient.

3.3.21   *MANTA ALFREDI*   (Krefft, 1868)

Phylum: Chordata   Subphylum: Vertebrata
Class: Chondrichthyes   Subclass: Elasmobranchii
Order: Myliobtiformes   Family: Myliobatidae

Common name: Alfred manta.

Geographical distribution: Tropical; Indo-West Pacific: Red Sea, South Africa, Thailand to Western Australia; Australia and Hawaiian Islands.

Habitat: Marine; benthopelagic; inshore, coral, and rocky reefs; atolls and bays.

Distinctive features: It is dorsoventrally flattened and has large, triangular pectoral fins on either side of the disk. At the front, it has a pair of cephalic fins which are forward extensions of the pectoral fins. These can be rolled up in a spiral for swimming or can be flared out to channel water into the large, forward-pointing, rectangular mouth when the animal is feeding. Eyes and the spiracles are on the side of the head behind the cephalic fins, and five gill slits are on the ventral surface. It has a small dorsal fin and the tail is long and whip-like. It does not have a spiny tail. Color of the dorsal side is dark black to midnight blue with scattered whitish and grayish areas on top head. Ventral surface is white, sometimes with dark spots and blotches. It has a maximum size of 3.5 m DW, maximum total length of 5.5 m and weight of 1.4 t (Marshall et al., 2009).

Food habits: This species is a filter feeder and eats large quantities of zooplankton, which it swallows with its open mouth as it swims.

Reproduction: Eggs of this species develop inside the female's body for about 1 year and then hatch internally. Each female gives birth to one or two pups. Newborns measure 1.5 m across and grows rapidly, almost doubling in size during the first year of life. Males mature when the disk-like body measures 2.5–3 m across and females at 3–3.9 m.

Predators: Predators of manta rays are large sharks, orcas (*Orcinus orca*), and false killer whales (*Pseudorca crassidens*).

Parasites: Not reported.

IUCN conservation status: Vulnerable.

3.3.22 *MANTA BIROSTRIS* (Walbaum, 1792)

Common name: Giant oceanic manta ray.

Geographical distribution: Circumglobal in tropical and temperate waters, this species has a widespread distribution.

Habitat: It is seen along productive coastlines with regular upwelling, oceanic island groups, and particularly offshore pinnacles and seamounts. It is also commonly encountered on shallow reefs, sandy bottom areas, and seagrass beds.

Distinctive features: It is dorsoventrally flattened and has large, triangular pectoral fins on either side of the disk. At the front, it has a pair of cephalic fins which are forward extensions of the pectoral fins. These can be rolled up in a spiral for swimming or can be flared out to channel water into the large, forward-pointing, rectangular mouth when the animal is feeding. Teeth are in a band of 18 rows and are restricted to the central part of the lower jaw. Eyes and the spiracles are on the side of the head behind the cephalic fins, and the gill slits are on the ventral surface. It has a small dorsal fin and the tail is long and whip like. It does not have a spiny tail but has a knob-like bulge at base of its tail. Skin is smooth with a scattering of conical and ridge-shaped tubercles. Coloration of the dorsal surface is black, dark brown, or steely blue, sometimes with a few pale spots and usually with a pale edge. Ventral surface is white, sometimes with dark spots and blotches. This species grows to a maximum size of 7 m DW and weight of 2 t. Estimated lifespan of this species is about 40 years.

Food habits: Feeds on tiny planktonic organisms by filtering large volumes of water through its mouth.

Reproduction: As in other rays, fertilization in this species is internal, with the male transferring its sperm to the female using a pair of "claspers" on the inner part of the pelvic fins. Developing eggs remain inside the female's body for about 1 year and then hatch internally. Female gives birth to one or occasionally two young. Newborn ray measures 1.2–1.5 m across. It matures at a DW of about 4–4.5 m in males and 5–5.5 m in females.

Predators: Predators of the giant manta ray are large sharks, orcas (*Orcinus orca*), and false killer whales (*Pseudorca crassidens*).

Parasites

Copepoda: *Anthosoma crassum*

Entepherus laminipes

IUCN conservation status: Vulnerable.

3.3.23 *MOBULA EREGOODOOTENKEE* (Bleeker, 1859)

Phylum: Chordata   Subphylum: Vertebrata
Class: Chondrichthyes   Subclass: Elasmobranchii
Order: Myliobtiformes   Family: Myliobatidae

Common name: Longhorned mobula, pygmy devilray.

Geographical distribution: Tropical; Indo-West Pacific: Red Sea, Arabian Sea, Persian Gulf to South Africa, Philippines, Viet Nam, northern Australia.

Habitat: Marine; pelagic, coastal, and oceanic waters.

Distinctive features: In this species, large pectoral fins are fused to the sides of its head and form a diamond shaped wing-like disk. Eyes are positioned on the side of the head and broad gill openings are situated underneath the front half of the pectoral fins. A thin, spineless tail projects from the rear of its flattened body. Extending forward from either side of its head are two prominent lobes which funnel planktonic food into the mouth on the underside of its head. Body coloration is brownish-gray above and whitish below. Underside of pectorals is with semicircular black blotch along middle of anterior edge. It grows to a maximum size of 1 m DW.

Food habits: Feeds on plankton and small fish.

Reproduction: It shows aplacental viviparity. Embryo develops in a membranous egg within the mother. After hatching, it remains in the mother and continues to be nourished by the yolk sac until it is ready to emerge. Each litter has just one offspring.

Predators: Not reported.

Parasites

Cestoda: *Hexacanalis govindi*

Hexacanalis yamagutii

Polypocephalus digholensis

Polypocephalus karbharii

IUCN conservation status: Near threatened.

3.3.24 *MOBULA HYPOSTOMA* (Bancroft, 1831)

Common name: Lesser devil ray.

Geographical distribution: Tropical; western Atlantic: New Jersey, USA to Santos; Brazil and Argentina; Eastern Atlantic: St. Paul's Rock.

Habitat: Marine; pelagic–neritic; shallow coastal waters.

Distinctive features: This species has forward facing, horn-like cephalic fins. Its distinctive large body disk is covered in small denticles. Tail is long, slender, and whip like without a spine. Head is relatively small and narrow with teeth present in the lower and upper jaws. Upper surface is black. Outer cephalic fins, lower parts of the disk, and the tail are a pale yellowish or grayish white, which continues along the ventral surface. disk of this species is about twice as wide as long, but it does not exceed 120 cm in DW.

Food habits: Feeds mainly on planktonic crustaceans and small schooling fishes

Reproduction: It shows aplacental viviparity. Female carries a single embryo, which is initially fed on yolk and subsequently by additional nourishment from greenish uterine milk. Females may mature at 107 cm in DW and newborn measures 55 cm DW.

Predators: Larger fish and marine mammals are potential predators of the Atlantic devil ray.

Parasites

Cestoda: *Rhinebothrium anterophallum*

IUCN conservation status: Data deficient.

3.3.25   *MOBULA JAPANICA* (Müller & Henle, 1841) = *Mobula diabolus*

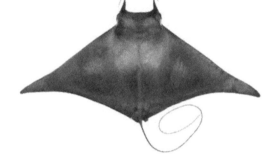

Common name: Spinetail mobula.

Geographical distribution: Subtropical; Indo-Pacific: South Africa, Arabian Sea Hawaiian Islands and Polynesia; Eastern Pacific: on the continental coast; Eastern Atlantic.

Habitat: Marine; reef-associated; inshore and oceanic waters.

Distinctive features: Body of this species is flattened into a disk which is much wider than it is long and is rhomboid in shape. It has a very long, whip-like tail which has a sting at the tip, a spine at the base, and a row of small white "teeth," known as denticles, along each side. Upperside of body is dark blue or black, with slit-like spiracles and white areas behind the eyes. Small denticles are also present on the upperside of the disk, as well as on the cephalic fins, lower jaw, gills, abdomen, and the underside of the pelvic fins. Maximum size and weight of this species are 317 cm DW and 115 kg, respectively.

Food habits: Feeds mainly on euphausiids (mainly *Nictiphanes simplex*) and to a lesser extent on copepods, crustacean larvae, and small fishes.

Reproduction: This species is ovoviviparous. The female has just one functional ovary and during the first stage of internal development, the embryo is initially enclosed within an egg and nutrients are gained from the yolk sac. After hatching from the egg inside the female, the embryo continues to develop and obtains its required nutrients from the fluid in the female's uterus until the pup is born live. Female gives birth to a single litter containing one pup. DW of a pup at birth is 85 cm.

Predators: Not reported.

Parasites

Cestoda: *Fellicocestus mobulae*

Healyum harenamica

Healyum pulvis

Hemionchos maior

Hemionchos mobulae

Hemionchos striatus

Quadcuspibothrium francis

Copepoda: *Entepherus laminipes*

IUCN conservation status: Near threatened.

3.3.26 MOBULA KUHLII (Müller & Henle, 1841)

Common name: Shortfin devil ray.

Geographical distribution: Tropical; Indo-West Pacific: eastern coast of Africa to Indonesia.

Habitat: Marine; pelagic; coastal and oceanic waters.

Distinctive features: It is flattened horizontally with a wide central disk and head is short with small cephalic fins. Large pectoral fins have curved tips and dorsal fin has a white tip. Tail lacks a spine and is shorter than the body. Dorsal surface is brown and does not bear any placoid scales. Ventral surface is white. It grows to a maximum size of 120 cm DW and a weight of 30 kg.

Food habits: Feeds on plankton, small fish, and squid.

Reproduction: This species shows ovoviparity (aplacental viviparity), with embryos feeding initially on yolk. Additional nourishment for the embryos is from the mother by indirect absorption of uterine fluid (enriched with mucus, fat, or protein) through specialized structures. A single, relatively large pup is produced per litter and the gestation period is 1–3 years. Males mature at 115–119 cm DW (Randall, 1995).

Predators: Not reported.

Parasites

Cestoda: *Crassuseptum pietrafacei*

Copepoda: *Caligus elongatus*

Eudactylina oliveri

IUCN conservation status: Data deficient.

3.3.27 *MOBULA MOBULAR* (Bonnaterre, 1788)

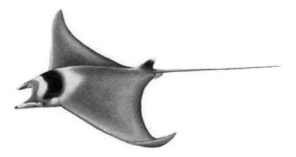

Common name: Devil fish, giant devil ray.

Geographical distribution: Subtropical; eastern Atlantic: southwest Ireland, Mediterranean Sea and Portugal south to Senegal, including the Canary and Azores islands.

Habitat: Marine; pelagic–neritic; oceanodromous; continental shelves and near oceanic islands.

Distinctive features: This species has large, pointed pectoral fins. Upper surface is dark and belly is mostly white. Cephalic fins which resemble devil horns are situated on the head and point forward and slightly down. Mouth is located on the underside. At the base of the slender tail, there is a serrated spine. Maximum total length and maximum DW of this species are 650 and 520 cm, respectively.

Food habits: Feeds on plankton and small pelagic fishes.

Reproduction: This species shows ovoviparity (aplacental viviparity), with embryos feeding initially on yolk. Additional nourishment for the embryos is from the mother by indirect absorption of uterine fluid (enriched with mucus, fat, or protein) through specialized structures. Gestation period is long, lasting about 25 months and females give birth to only one, or rarely two, pups. The newborn devil rays may measure up to 180 cm in DW.

Predators: Not reported.

Parasites: Not reported.

IUCN conservation status: Endangered.

3.3.28 *MOBULA MUNKIANA,* Notarbartolo-di-Sciara, 1987

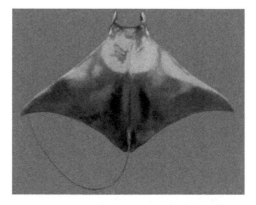

Common name: Munk's devil ray.

Geographical distribution: Tropical; eastern Pacific: Gulf of California, Ecuador, and Galapagos Islands.

Habitat: Marine; pelagic–oceanic; oceanodromous; coastal and oceanic waters, but also found near the bottom.

Distinctive features: This species has long, pointed, pectoral fins which it strikes up and down, like wings, for propulsion. On either side of the head, and in front of its prominent eyes, there are two fleshy lobes which project forward to funnel food into the mouth. Dorsal fin is small, and tail is long, flattened and spineless. Upper surface of body is dark purplish to mauve-gray, while belly is white. Toward wing-tips, coloration is becoming blue-gray. This species grows to a maximum size of 220 cm DW.

Food habits: Feeds mainly on planktonic crustaceans, but also takes small schooling fishes.

Reproduction: This species shows ovoviparity (aplacental viviparity), with embryos feeding initially on yolk. Additional nourishment for the embryos is from the mother by indirect absorption of uterine fluid (enriched with mucus, fat, or protein) through specialized structures. Female produces just a single pup per litter.

Predators: Not reported.

Parasites

Cestoda: *Hemionchos mobulae*

IUCN conservation status: Near threatened.

3.3.29 *MOBULA ROCHEBRUNEI* (Vaillant, 1879)

Image not available

Common name: Lesser Guinean devil ray.

Geographical distribution: Tropical; Eastern Atlantic: Mauritania to Angola; Brazil.

Habitat: Marine; pelagic–neritic.

Distinctive features: Head of this species is less than 20% of DW. Mouth is subterminal in position. Teeth are present on both jaws. Tooth band of lower jaw is less than 50% of mouth width. Cephalic fins which resemble horns serve as a funnel to bring food near the mouth. This species has a maximum size of 133 cm DW.

Food habits: Feeds on small fish and plankton.

Reproduction: This species shows ovoviparity (aplacental viviparity), with embryos feeding initially on yolk. Additional nourishment for the embryos is from the mother by indirect absorption of uterine fluid (enriched with mucus, fat, or protein) through specialized structures. Female gives birth to one young per litter.

Predators: Not reported.

Parasites

Copepoda: *Entepherus laminipes*

IUCN conservation status: Vulnerable

3.3.30 *MOBULA TARAPACANA* (Philippi, 1892)

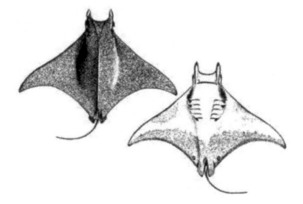

Dorsal and Ventral views

Common name: Chilean devil ray.

Geographical distribution: Tropical; Circumtropical; western Atlantic: Venezuela; eastern Atlantic: Côte d'Ivoire and South Africa; Cape Verde; western Indian Ocean: north-western Red Sea. Western Pacific: Japan, Taiwan, and Australia; Eastern Pacific: Gulf of California and Chile.

Habitat: Marine; reef-associated; oceanodromous; oceanic and coastal waters.

Distinctive features: It is a large species with a long head bearing short head fins. Dorsal fin is plain and pectoral fins are with strongly curved, swept-back tips. Upper disk is densely covered with small, pointed denticles. Tail which is shorter than the disk is without spine. Coloration is dark blue or olive-green to brownish above. Ventral side is white anteriorly, and gray posteriorly, with an irregular but distinct line of demarcation. There is no caudal fin. Maximum size and weight of this species are 328 cm DW and 350 kg, respectively (McEachran & di Sciar, 1995).

Food habits: Feeds on small fishes and planktonic crustaceans.

Reproduction: This species shows ovoviparity (aplacental viviparity), with embryos feeding initially on yolk. Additional nourishment for the embryos is from the mother by indirect absorption of uterine fluid (enriched with mucus, fat, or protein) through specialized structures.

Predators: Not reported.

Parasites

Copepoda: *Entepherus laminipes*

IUCN conservation status: Data deficient.

3.3.31   *MOBULA THURSTONI* (Lloyd, 1908)

Dorsal and ventral views

Common name: Smoothtail mobula.

Geographical distribution: Subtropical; probably circumtropical but in scattered localities; eastern Atlantic: off Senegal and Côte d'Ivoire. Indian Ocean: off South Africa, Bay of Bengal, and probably Indonesia; western Pacific: Gulf of Thailand and northeastern Australia; eastern Pacific: southern California, USA to Costa Rica, including the Gulf of Tehuantepec; reported from Chile.

Habitat: Marine; pelagic; coastal and oceanic waters.

Distinctive features: This species has short head fins. Dorsal fin is white-tipped and pectoral fins are with swept-back tips. Upper disk is sparsely covered with small, blunt denticles. Tail which is shorter than the disk is without spine. Coloration is dark blue to black above and white below with silvery pectoral fin tips. There is no caudal fin. It grows to a maximum size of 220 cm DW (Last & Stevens, 1994).

Food habits: Feeds mainly on planktonic crustaceans (mostly small shrimp-like animals).

Reproduction: This species shows ovoviparity (aplacental viviparity), with embryos feeding initially on yolk. Additional nourishment for the embryos is from the mother by indirect absorption of uterine fluid (enriched with mucus, fat, or protein) through specialized structures.

Predators: Not reported.

Parasites

Cestoda: *Hemionchos striatus*

Mobulocestus lepidoscolex

Mobulocestus mollis

Mobulocestus nephritidis

Trematoda: *Nagmia cisloi*

Copepoda: *Echthrogaleus disciarai*

Entepherus laminipes

IUCN conservation status: Near threatened.

3.3.32 *PTEROMYLAEUS ASPERRIMUS* (Gilbert, 1898)

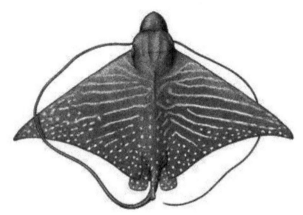

Phylum: Chordata   Subphylum: Vertebrata
Class: Chondrichthyes   Subclass: Elasmobranchii
Order: Myliobtiformes   Family: Myliobatidae.

Common name: Rough eagle ray.

Geographical distribution: Tropical; eastern Pacific: Panama and the Galapagos Islands.

Habitat: Marine; demersal; soft bottoms in coastal waters.

Distinctive features: Upper surface of this species is with 8–10 narrow transverse bars anteriorly. Most of these bars become spot-like at the outer margin of the disk. A series of irregular-sized spots are seen near wing tips and along posterior margins of wings extending forward on midline body. Disk of this species is twice as broad as long. Anterior margins are convex and posterior margins are concave. Pectoral wings are pointed and snout projects in a single flexible lobe. Spiracles are large. Tail is long, slender and is whip like. It is more than three times the length of body with a spine behind single dorsal fin. Spine is long, twice the length of dorsal fin. This species grows to a maximum size of 79.0 cm DW (Grove & Lavenberg, 1997).

Food habits: Feeds on small fish, crabs, shrimp, lobsters, cephalopods and polychaete worms.

Reproduction: It is presumably aplacental viviparous, like other eagle rays, but virtually no information is available on this species' biology or life-history parameters (Valenti & Kyne, 2009).

Predators: Not reported.

Parasites: Not reported.

IUCN conservation status: Data deficient.

3.3.33 *PTEROMYLAEUS BOVINUS* (Geoffroy Saint-Hilaire, 1817)

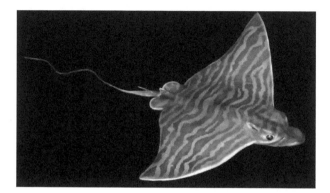

Common name: Bull ray.

Geographical distribution: Tropical, subtropical, and warm temperate waters; Eastern Atlantic: Portugal and Morocco to Angola, including the Mediterranean, Madeira, and the Canary Islands; from Saldanha Bay to Natal (South Africa) and southern Mozambique.

Habitat: Marine; brackish; benthopelagic; coastal and offshore.

Distinctive features: This species has been named due to the shape of its head. It is sometimes called the duckbill ray for its long, flat, round snout. Females are larger and heavier than males. It has a tail spine averaging 6.1 cm in length in females and 3.2 cm in males. Values of its maximum total length, width and weight are 2.5 m, 1.8 m, and 100 kg, respectively.

Food habits: Feeds on invertebrates including crabs, hermit crabs, squids, prawns, gastropod molluscs and bivalve molluscs.

Reproduction: This species shows ovoviparity (aplacental viviparity), with embryos feeding initially on yolk. Additional nourishment for the embryos is from the mother by indirect absorption of uterine fluid (enriched with mucus, fat, or protein) through specialized structure. Gestation period is believed to be about 1 year and 3–4 young may be carried at one time by a female.

Predators: Not reported.

Parasites: Not reported.

IUCN conservation status: Data deficient.

3.4   COWNOSE STINGRAYS (RHINOPTERIDAE)

3.4.1   *RHINOPTERA ADSPERSA,* Müller & Henle, 1841

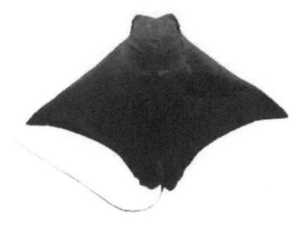

Phylum: Chordata   Subphylum: Vertebrata
Class: Chondrichthyes   Subclass: Elasmobranchii
Order: Myliobtiformes   Family: Rhinopteridae

Common name: Rough cownose ray.

Geographical distribution: Tropical; Indo-West Pacific: off India, Malaysia, and East Indies.

Habitat: Marine; benthopelagic.

Distinctive features: This is a rare species with a rough back with small stellate based spines. Upper teeth are in nine rows and lower teeth in seven rows. Body is greenish brown above and lighter below. Maximum total length of this species is 99 cm (Munro, 2000).

Food habits: Feeds on invertebrates including crabs, hermit crabs, squids, prawns, gastropod molluscs, and bivalve molluscs.

Reproduction: This species exhibits ovoviparity (aplacental viviparity), with embryos feeding initially on yolk. Additional nourishment for the embryos is from the mother by indirect absorption of uterine fluid (enriched with mucus, fat, or protein) through specialized structures.

Predators: Not reported.

Parasites

Cestoda: *Eniochobothrium qatarense*

IUCN conservation status: Not evaluated.

3.4.2   *RHINOPTERA BONASUS* (Mitchill, 1815)

Common name: Cownose ray.

Geographical distribution: Eastern Atlantic Ocean including Mauritania, Senegal, and Guinea; western Atlantic: from southern New England to northern Florida (USA); Gulf of Mexico, Trinidad, Venezuela, and Brazil.

Habitat: Inshore waters; backishwaters.

Distinctive features: disk of this species is 1.7 times as broad as it is long. Eyes and the spiracles of this species are located on the sides of the broad head. Pectorals arise from the sides of the head. Outer corners of the pectorals are pointed and are concave toward their posterior margins. Dorsal fin originates approximately opposite the rear ends of the bases of the pelvic fins and is rounded above. Tail which is round to oval in cross-section is moderately stout near the anterior spine, and narrows rearward, tapering to a lash-like tip. Length of the tail is about twice as long as the body. There are one or two tail spines. The first spine (posterior spine) is located directly behind the base of the dorsal fin. When present, second spine (anterior spine) has a free portion which is about half as long as the anterior margin of the pelvic fin. Anterior spine varies in length from very short to as long as the posterior spine. Spines have marginal teeth with broad bases and sharp tips. Number of teeth ranges from 22 to 45. Dorsal surface of this species is light to dark brown, and it may sometimes have a yellowish tint. Ventral surface is white or yellowish white. Tail coloration is very much similar to that of body. Maximum size of this species is 213 cm DW (FLMNH Ichthyology Dept.).

Food habits: Diet consists primarily of bivalve molluscs. Other common prey items include nekton, zoobenthos, finfish, benthos crustaceans, mollusks, bony fish, crabs, lobsters, bivalves, and gastropods.

Reproduction: Like other elasmobranchs, this species is also ovoviviparous and gestation period is believed to be 11–12 months. Each female may give birth to one to six pups. Males and female mature at a DW of 79 cm and 61 cm, respectively.

Predators: Not reported.

Parasites

Cestoda: *Dioecotaenia cancellata*

Duplicibothrium minutum

Echinobothrium bonasum

Glyphobothrium zwerneri

Mecistobothrium brevispine

Nybelinia sp.

Rhinoptericola megacantha

Rhodobothrium paucitesticulare

Tylocephalum bonasum

Tylocephalum brooksi

Tylocephalum marsupium

Tylocephalum pingue

Tylocephalum sp.

Zygorhynchus sp.

IUCN conservation status: Near threatened.

3.4.3 *RHINOPTERA BRASILIENSIS,* Müller, 1836

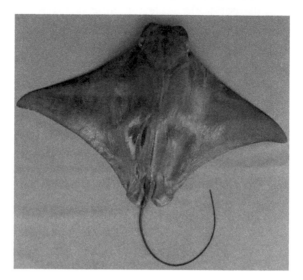

Common name: Brazilian cownose ray.

Geographical distribution: Tropical; southwest Atlantic.

Habitat: Marine; demersal; shallow coastal waters.

Distinctive features: This species has a broader mouth with three central rows of broad teeth. Median row of teeth is wider transversely to their lengths. Upper jaw has nine series of teeth. Males of this species range in size from 78 to 91 cm DW with a brown back and white or light yellow belly. Females are larger, ranging from 77 to 102 cm DW with similar colorations. This species grows to a maximum size of 90.0 cm DW.

Food habits: Feeds primarily on molluscs.

Reproduction: This species shows ovoviparity (aplacental viviparity), with embryos feeding initially on yolk. Additional nourishment for the embryos is from the mother by indirect absorption of uterine fluid (enriched with mucus, fat, or protein) through specialized structures. Breeding cycle of females may be biennial, although specific information is not available. Female carries only one embryo at a time. Size at birth is 43–48 cm DW.

Predators: Not reported.

Parasites: Not known.

IUCN conservation status: Endangered.

3.4.4 *RHINOPTERA JAVANICA,* Müller & Henle, 1841

Common name: Flapnose ray.

Geographical distribution: Tropical; Indo-West Pacific: Durban, South Africa, India, Thailand, Indonesia, southern China, Okinawa, Ryukyu Islands, and Australia.

Habitat: Marine; brackishwaters; bays, estuaries, coral reefs; over sand and mud bottoms; reef-associated.

Distinctive features: This species has an unusual-looking head, which features a double-lobed snout and indented forehead. Body is flattened, with the pectoral fins broadly expanded and fused with the head and trunk to form a disk. This smooth-skinned species is characterized by a kite-shaped body-disk which is brown on the upper surface and white below. Jaws are usually with seven rows of plate-like teeth and there is no caudal fin. Long, thin, whip-like tail of this species is distinctly demarcated from the body and is armed with one or more stings. Average total spine length of males is 52 mm and 112 mm in females. Total number of serrations in these sexes is 36 and 72, respectively. It grows to a maximum size of 150 cm DW and weight of 4.5 kg (Schwartz, 2007).

Food habits: Feeds on a diet of clams, oysters, and crustaceans.

Reproduction: This species shows ovoviparity (aplacental viviparity), with embryos feeding initially on yolk. Additional nourishment for the embryos from the mother. It attains 150 cm DM with the largest recorded embryo 63 cm. It has one to two young per litter due to its large size at birth (Dudley et al., 2006).

Predators: Not reported.

Parasites

Cestoda: *Echinobothrium rhinoptera*

Eniochobothrium gracile

Phyllobothrium rhinoptera

Tetrarhynchobothrium unionifactor

IUCN conservation status: Vulnerable.

3.4.5   RHINOPTERA JAYAKARI, Boulenger, 1895

Common name: Oman cownose ray.

Geographical distribution: Tropical; western Indian Ocean: known only from Oman.

Habitat: Marine; benthopelagic.

Distinctive features: Each anterior margin of disk is nearly straight and posterior margin is concave. Disk tip is pointed. Upper part of snout is bilobed. Tail is short and is only two-thirds disk length. A serrate spine is present on tail behind dorsal fin. Body coloration is olivaceous above and white below. Tail is black distally. Total length and DW are 1.3 m and 90 cm, respectively (Randall, 1995).

Food habits: Feeds on a diet of clams, oysters, and crustacea.

Reproduction: This species shows ovoviparity (aplacental viviparity) with embryos feeding initially on yolk, subsequently receiving additional nourishment from the mother by indirect absorption of uterine fluid (enriched with mucus, fat, or protein) through specialized structures.

Predators: Not reported.

Parasites: Not known.

IUCN conservation status: Not evaluated.

3.4.6 *RHINOPTERA MARGINATA* (Geoffroy Saint-Hilaire, 1817)

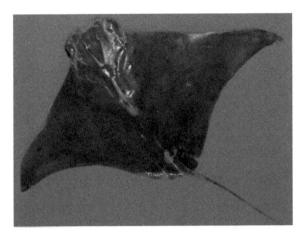

Common name: Lusitanian cownose ray.

Geographical distribution: Subtropical and warm temperate regions; Eastern Atlantic: southern Spain to Senegal, including the Mediterranean.

Habitat: Marine; benthopelagic; coastal waters; sandy-muddy bottoms.

Distinctive features: Front of head of this species is distinctly concave. Snout is notched. Teeth are seen in nine rows in upper jaw and seven in lower. Middle row is not more than three times as broad as long and outer rows are pentagonal. Pectoral fins are slightly falcate and their outer angle is blunt. Pelvic fins are longer than wide. Upper surfaces are without thorns or thornlets. Tail is whip like with one sting and a small dorsal fin at base. Upper surfaces are greenish-brown to bronze and underside is whitish. This species grows to a maximum size of 2 m DM.

Food habits: Feeds on bottom-living molluscs, crustaceans, and fishes.

Reproduction: This species shows ovoviparity (aplacental viviparity), with embryos feeding initially on yolk. Additional nourishment for the embryos is from the mother by indirect absorption of uterine fluid (enriched with mucus, fat, or protein) through specialized structures. Males mature at 77 cm in length and females over 80 cm. Gestation period is up to 1 year and each female gives birth to 2–6 pups.

Predators: Not reported.

Parasites: Not known.

IUCN conservation status: Near threatened.

3.4.7 *RHINOPTERA NEGLECTA,* Ogilby, 1912

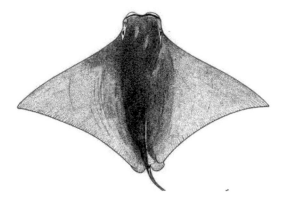

Common name: Australian cownose ray.

Geographical distribution: Subtropical; western Pacific: Queensland to New South Wales in Australia.

Habitat: Marine; reef-associated; coastal waters and often enter estuaries.

Distinctive features: This species has an unusual bilobed head with two large fleshy lobes under the snout. Coloration is dark grayish above and is white below. There is a single dorsal fin and one or more serrated near the base of the whip-like tail. This species grows to a maximum size of 86 cm DW.

Food habits: The flat, plate-like teeth are used by this species to crush and grind crustaceans and other invertebrates.

Reproduction: This specie shows ovoviparity (aplacental viviparity), with embryos feeding initially on yolk. Additional nourishment for the embryos is from the mother by indirect absorption of uterine fluid (enriched with mucus, fat, or protein) through specialized structures.

Predators: Not reported.

Parasites: Monogenea: *Euzetia occultum*

IUCN conservation status: Data deficient.

3.4.8 *RHINOPTERA STEINDACHNERI*, Evermann & Jenkins, 1891

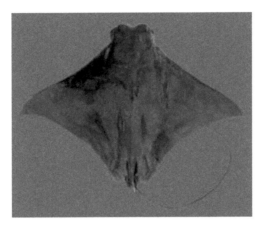

Common name: Pacific cownose ray, golden cownose ray.

Geographical distribution: Tropical; Eastern Pacific: Gulf of California to Costa Rica and Galapagos Islands.

Habitat: Marine; oceanodromous; soft bottoms, near rocky or coral reefs; also near reef drop-offs.

Distinctive features: Tail of this species is whip like. Plate-like teeth are seen in the mouth. Back coloration of this species is golden brown to brown and grayish brown. Belly is white. This species grows to a maximum size of 104 cm DW.

Food habits: Feeds mainly on molluscs.

Reproduction: This species shows ovoviparity (aplacental viviparity), with embryos feeding initially on yolk. Additional nourishment for the embryos is from the mother by indirect absorption of uterine fluid (enriched with mucus, fat, or protein) through specialized structures. Only a single ovary is functional. Median size at maturity for females is 70.2 cm DW and for males is 69.9 cm DW. Fecundity is one offspring per female and gestation period is 10–12 months. Size at birth has been estimated at 38–45 cm DW after a gestation period of approximately 11–12 months.

Predators: Not reported.

Parasites

Cestoda: *Duplicibothrium cairae*

Duplicibothrium paulum

Echinobothrium fautleyae

Serendip deborahae

IUCN conservation status: Near threatened.

3.5   STINGREES OR ROUND STINGRAYS (UROLOPHIDAE)

3.5.1   *TRYGONOPTERA GALBA,* Last & Yearsley, 2008

Phylum: Chordata   Subphylum: Vertebrata
Class: Chondrichthyes   Subclass: Elasmobranchii
Order: Myliobtiformes   Family: Urolophidae

Common name: Yellow shovelnose stingaree.

Geographical distribution: Subtropical; eastern Indian Ocean: Australia.

Habitat: Benthopelagic; prefers sandy habitats.

Distinctive features: This species has an oval pectoral fin disk which is slightly wider than long. Anterior margins of the disk are weakly convex and are converging on a fairly elongated, fleshy snout. Tip of the snout does not protrude past the disk. Eyes are medium-sized, well spaced and are slightly elevated. Spiracles behind the eyes are comma shaped. Posterior margins of the spiracles are angular. Mouth is small and lower jaw conceals upper jaw and bears a prominent, corrugated patch of papillae. Small teeth have oval to diamond-shaped bases and are arranged in a quincunx pattern. There are 19–20 upper and 22–23 lower tooth rows. Floor of the mouth bears eight or more papillae. There are five pairs of gill slits which are S-shaped. Pelvic fins are roughly triangular. Males have short, thick claspers. Tail measures less than an eighth as long as the disk and is moderately flattened at the base, tapering smoothly to a lance-shaped caudal fin. There are no dorsal fins or fin folds. A single, serrated stinging spine is located atop the tail. Skin is devoid of dermal denticles. Upper surface of the disk and tail is a deep, even yellow to yellowish brown in color, becoming darker on the caudal fin. Belly is white to yellowish. This species grows to maximum total length of 32.8 cm only.

Food habits: Biology of this species is unknown.

Reproduction: It is presumably aplacental viviparous like other stingrays. Newborns measure 16 cm long and males reach sexual maturity at 33–36 cm long (Last & Marshall, 2009).

Predators: Not reported.

Parasites: Not known.

IUCN conservation status: Data deficient.

3.5.2   *TRYGONOPTERA IMITATA,* Yearsley et al., 2008

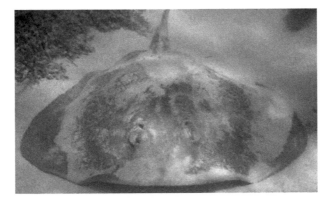

Common name: Eastern shovelnose stingaree.

Geographical distribution: Temperate; southern Oceans: Australia.

Habitat: Marine; benthopelagic; freshwater; near Beaches and coastal waters; inshore bays.

Distinctive features: The pectoral fin disk of this species is rounded in shape, wider than long, and thick at the center. Anterior margins of the disk are straight to gently convex, and converge on the fleshy, non-protruding snout. Outer corners of the disk are broadly rounded. Medium-sized eyes are followed by larger comma-shaped spiracles with angular posterior rim. A small mouth overhangs. Lower jaw bears three papillae in the middle and one or two papillae near either corner. Teeth are blunt with oval bases. They are arranged in a quincunx pattern, numbering around 22 upper and 24 lower rows. Five pairs of gill slits are S-shaped. Pelvic fins are modestly sized and roughly triangular. Males have robust, tapering claspers. Tail ends in a lance-like caudal fin. It measures around three-quarters as long as the disk. Usually, two serrated stinging spines, with the upper generally larger than the lower, are present atop the tail. Tail lacks dorsal fins and fin folds. Skin is completely smooth. Dorsal coloration ranges from yellowish to dark grayish brown, which is darkest toward the midline and lightest toward the fin margins. Underside is light colored with a dark fin margins and some-times irregular dark blotches on the abdomen. Tail is entirely dark past the base. This species reaches a maximum known total length of 80 cm.

Food habits: Feeds on small benthic organisms, primarily polychaete worms.

Reproduction: Females have a single functional ovary, on the right, and are aplacental viviparous like other stingray species. Once fertilized, the fairly large eggs are retained in the uterus without developing for 5–8 months, after which the embryos emerge and rapidly develop to term over a period of 4–7 months. Embryos are at first sustained by internal and external yolk sacs; once the yolk supply is exhausted, they are provided with nutrient-rich uterine milk by the mother. Females bear one litter per year and litter size is up to seven. Newborns measure 20–25 cm long with a weight of 150 g. Males attain sexual maturity at about 46 cm long and 4.5 years of age, and females at 49 cm long and 5 years of age. Maximum life span is at least 10 years for males and 12 years for females.

Predators: Not reported.

Parasites: Not known.

IUCN conservation status: Near threatened.

3.5.3 *TRYGONOPTERA MUCOSA* (Whitley, 1939)

Common name: Western shovelnose stingaree.

Geographical distribution: Temperate; eastern Indian Ocean: endemic to Australia.

Habitat: Marine; demersal; sandy substrates or near seagrass.

Distinctive features: This species has a rounded pectoral fin disk which is slightly wider than long. Its leading margins are nearly straight and converge on the fleshy snout which has a blunt, non-protruding tip. Modestly sized eyes are followed by much larger, comma-shaped spiracles. Mouth is small. Lower jaw bears a patch of papillae (nipple-shaped structures). There are 7–9 small papillae across the floor of the mouth. Teeth are small with roughly oval bases. Five pairs of gill slits are short. Pelvic fins are small and rounded. Tail terminates in a low leaf-shaped caudal fin. Tail bears a serrated stinging spine on the dorsal surface about midway along its length. There are no dorsal fins or lateral fin folds. Skin is completely devoid of dermal denticles. This species is grayish to dark brown in color above and is becoming dusky to black on the caudal fin. Belly is pale brownish or yellowish. This species grows to a maximum total length of 44 cm only. Maximum reported DW is 28 cm for males and 37 cm for females.

Food habits: Feeds predominantly on polychaete worms, crustaceans (including shrimp, amphipods, isopods, mysids, and tanaids) and sipunculid worms constitute minor secondary food sources and molluscs, echinoderms, and bony fishes are very rarely eaten.

Reproduction: Like other stingrays, this species is aplacental viviparous. Adult females have a single functional uterus (on the left), and produce litters of usually one, but rarely two pups every year. Total gestation period lasts about a year; once the developing embryos exhaust their supply of yolk, they are supplied with nutrient-rich uterine milk by the mother. Newly born rays measure 11 cm across. Sexual maturity is attained at 22 cm across and 2 years of age for males, and 25 cm across and 5 years of age for females. Maximum lifespan is 12 years for males and 17 years for females.

Predators: Not reported.

Parasites: Nematoda: *Echinocephalus overstreeti*

IUCN conservation status: Least concern.

3.5.4   *TRYGONOPTERA OVALIS,* Last & Gomon, 1987

Common name: Striped stingaree.

Geographical distribution: Subtropical; eastern Indian Ocean: Western Australia.

Habitat: Marine; near reefs, sea grass beds and sandy beaches.

Distinctive features: Pectoral fin disk of this species is oval in shape. It is slightly longer than wide with the anterior margins converging on the rounded, non-protruding tip of the fleshy snout. Medium-sized eyes are immediately followed by comma-shaped spiracles with angular posterior rims. It has a small mouth. There are numerous papillae on the lower jaw, while four tiny papillae are found on the floor of the mouth. Small teeth have roughly oval bases. Five pairs of gill slits are short. Pelvic fins are small and rounded. Tail measures 75–100% as long as the disk. It is slightly flattened at the base and terminates in a fairly large, deep, leaf-shaped caudal fin. Upper surface of the tail mounts a serrated stinging spine, which is immediately preceded by a small dorsal fin. Skin is entirely smooth. Dorsal coloration of this species is grayish to grayish brown. There is a dark mask-like pattern around the eyes that may extend to the tip of the snout as well as a pair of dark blotches at the center of the disk. These blotches are drawn out posteriorly into stripes that run onto the tail. Belly is pale. Margins of the fins are dark. Caudal fin is gray or black with a darker trailing margin. Maximum known total length is 61 cm.

Food habits: Feeds on polychaete worms, crustaceans and molluscs.

Reproduction: Reproduction is presumably aplacental viviparous like other stingrays. Males attain sexual maturity at 35 cm long. Little is known of its natural history.

Predators: Not reported.

Parasites: Not known.

IUCN conservation status: Least concern.

3.5.5 TRYGONOPTERA PERSONATA, Last & Gomon, 1987

Common name: Masked stingaree.

Geographical distribution: Subtropical; eastern Indian Ocean: Western Australia.

Habitat: Marine; demersal; favors offshore waters, sandy flats, and seagrass beds.

Distinctive features: Pectoral fin disk of this species as long as wide and has a rounded shape. Anterior margins of the disk are nearly straight and converge on the fleshy, non-protruding snout. Eyes are modestly sized and are immediately followed by comma-shaped spiracles with angular posterior rims. Mouth is small. Lower jaw bears a patch of subtle papillae and there are also 3–4 papillae on the floor of the month. Small teeth have roughly oval bases. Five pairs of gill slits are short. Pelvic fins are small with rounded margins. Tail is slightly flattened at the base. There is a single serrated stinging spine on the upper surface of the tail which is immediately preceded by a rather large dorsal fin. A leaf-shaped caudal fin is present and lateral fin folds are absent. Skin is devoid of dermal denticles. This species has pale to gray dorsal coloration with two large, distinctive dark blotches. Of these blotches, one is forming a mask around the eyes and the other is at the center of the disk. Belly is white. Dorsal and caudal fins are black in young rays, and gray in adults. This species has a maximum total length of only 47 cm. Males and females normally grow up to 27 cm and 31 cm across, respectively.

Food habits: Polychaete worms and crustaceans are the predominant sources and sipunculid worms, molluscs, and echinoderms are also occasionally consumed.

Reproduction: This species is aplacental viviparous. Females have a single functional uterus (the left) and produce a single pup (rarely two) per year. Embryos are sustained by nutrient-rich uterine milk produced by the mother. Total gestation period is 10–12 months. Males mature sexually at 22 cm across and females at 23 cm across. Both sexes mature on average at 4 years of age. Maximum lifespan is 10 years for males and 16 years for females.

Predators: Not reported.

Parasites: Not known.

IUCN conservation status: Least concern.

3.5.6 *TRYGONOPTERA TESTACEUS,* Müller & Henle, 1841 = *Urolophus testaceus*

Common name: Common stingaree.

Geographical distribution: Temperate; western Pacific: southern Queensland to New South Wales.

Habitat: Marine; brackish; demersal; sandy beaches and reefs; offshore; upstream in estuaries.

Distinctive features: This species has a small dorsal fin or a narrow ridge of skin in front of one or two strong spines on the tail. A leaf-shaped caudal fin is present. This species is dark brown to gray above and white below. This species grows to a maximum size of 47 cm in total length (http://australianmuseum. net.au/common- stingaree-trygonoptera-testacea-muller-henle-1841).

Food habits: Feeds on worms, crustaceans, bony fishes, and other stingarees.

Reproduction: This species is ovoviviparous like other stingrays. Males mature at a DW of 22 cm, while females reach sexual maturity at 26 cm DW. Only the left uterus and ovary are found to be functional in females.

Predators: Not reported.

Parasites: Not known.

IUCN conservation status: Near threatened.

3.5.7   *UROBATIS TUMBESENSIS* (Chirichigno & McEachran, 1979)
= Urolophus tumbesensis

Phylum: Chordata	Subphylum: Vertebrata
Class: Chondrichthyes	Subclass: Elasmobranchii
Order: Myliobtiformes	Family: Urolophidae

Common name: Tumbes round stingray.

Geographical distribution: Tropical; southeast Pacific: off Peru.

Habitat: Marine; demersal; shallow inshore habitats; estuarine waters; nearmangroves.

Distinctive features: This species has a rounded pectoral fin disk which is slightly wider than it is long. Tail is stout, bearing a serrated stinging spine, and terminates in a rounded caudal fin. Pelvic fins have abruptly rounded tips. Teeth have narrowly oval bases and no elevated cusps. Dorsal surface is covered uniformly by dermal denticles which are becoming larger toward the midline of the disk. There are also thorns on the dorsal surface of the disk and tail. Dorsal coloration consists of pale vermiculations separating brownish-white oval or circular spots about the size of the eye. Coloration is becoming more distinct toward the margin of the disk and on the pelvic fins. Denticles and tail spine are pale and belly is light tan with a dark border along the edge of the disk. This species grows to a maximum total length of 41 cm and weight of 872 g (Mejia-Falla & Navia, 2009).

Food habits: Feeds mainly on cladocerans and polychaetes.

Reproduction: It is ovoviviparous and males mature at 30 cm total length. No details are available on the biology of the species (Kyne & Valenti, 2007).

Predators: Not reported.

Parasites

Cestoda: *Pararhinebothroides hobergi*

IUCN conservation status: Data deficient.

3.5.8   *UROLOPHUS ARMATUS,* Müller & Henle, 1841

Phylum: Chordata	Subphylum: Vertebrata
Class: Chondrichthyes	Subclass: Elasmobranchii
Order: Myliobtiformes	Family: Urolophidae

Common name: New Ireland stingaree.

Geographical distribution: Tropical; western central Pacific: known only from New Ireland (Bismarck Archipelago), possibly New Guinea.

Habitat: Marine; demersal.

Distinctive features: Disk of this species is oval in shape and is wider than long with evenly rounded outer margins. Leading margins of the disk become slightly concave as they converge on the protruding, pointed snout. Eyes are tiny and are immediately followed by slightly larger spiracles. Small, bow-shaped mouth contains a single central papilla on the floor. Several tiny papillae are also scattered outside the lower jaw and on the tongue. There are 24 lower tooth rows. Pelvic fins are short and broad, with angular tips. Slender tail measures 92% as long as the disk and tapers evenly to a long, low leaf-shaped caudal fin. There are subtle skin folds running along either side. There is no dorsal fin. Two serrated stinging spines are placed atop the tail, about halfway along its length. This species is the only member of its family with dermal denticles. There are four rows of small, sharp, well-spaced thorns running along the middle of the back before the stings. Coloration is brown above with dark spots either scattered or merged into blotches and white below. This species grows to a maximum total length of only 17.4 cm.

Food habits: Nothing is known of habitat, depth distribution, or biology of this species (Last & Marshall, 2006c).

Reproduction: Virtually nothing is known of the natural history of this species. It is presumably aplacental viviparous with a small litter size, like other stingarees. It is likely to have low fecundity (one to two young/year) as with other urolophid species.

Predators: Not reported.

Parasites: Not known.

IUCN conservation status: Data deficient.

3.5.9 *UROLOPHUS AURANTIACUS,* Müller & Henle, 1841

Common name: Sepia stingray, sepia stingaree.

Geographical distribution: Temperate; northwest Pacific: Japan and the Ryukyu Islands to the East China Sea.

Habitat: Marine; demersal; sand and rock bottoms.

Distinctive features: Disk of this species is wider than long. Back is smooth. Snout is short and eyes are small. Tail is thick and short. While there is a rounded caudal fin, dorsal fin is absent. Average STL of females is 42 mm and total number of serrations is 41. Dorsal coloration is dark brown with brown spots and circles and belly is whit. Eyes are golden in color. It grows to a maximum size of 40 cm total length (Schwartz, 2007.

Food habits: Feeds on crustaceans, polychaete worms, and fish.

Reproduction: Like other stingrays, this species is ovoviviparous and its litter size is 1–4 and its gestation period is about 1 year (Last & Marshall, 2006a).

Predators: Not reported.

Parasites: Not known.

IUCN conservation status: Near threatened.

3.5.10 *UROLOPHUS BUCCULENTUS*, Macleay, 1884

Common name: Sandyback stingaree.

Geographical distribution: Temperate; eastern Indian Ocean to western Pacific: endemic to Australia.

Habitat: Marine; demersal; offshore bottom; outer continental shelf and uppermost slope.

Distinctive features: This species has a diamond-shaped pectoral fin disk which is much wider than long, with rounded outer corners and straight leading margins. Snout is fleshy and is slightly protruding at the tip. Small eyes are closely followed by comma-shaped spiracles with angular to rounded posterior rims. Mouth is fairly large and contains small teeth with roughly oval bases, as well as 14–16 papillae on the floor and a narrow patch of papillae on the lower jaw. Five pairs of gill slits are short. Pelvic fins are small with rounded margins. Tail is short and is strongly flattened with a skin fold running along each side. Upper surface of the tail bears a serrated stinging spine which is preceded by a relatively large dorsal fin. Caudal fin is lance like, short, and deep. Skin is devoid of dermal denticles. Coloration of this species is yellowish to brownish above. Dorsal and caudal fins are darker in juveniles and may be mottled in adults. Belly is plain white. It grows to a maximum size of 89 cm in total length.

Food habits: Preys primarily on crustaceans.

Reproduction: Reproduction is aplacental viviparous, with the developing embryos sustained by maternally produced uterine milk-like in other stingrays. Females bear litters of 1–5 pups every other year after a gestation period of 14–19 months. Newborn rays measure 17 cm long. Males attain sexual maturity at around 40 cm long and females at 50 cm long.

Predators: Not reported.

Parasites

Monogenea: *Calicotyle urolophi*

Cestoda: *Acanthobothrium robertsoni*

IUCN conservation status: Endangered.

3.5.11 *UROLOPHUS CIRCULARIS*, McKay, 1966

Common name: Circular stingaree.

Geographical distribution: Restricted range; off southwestern Australia between Esperance and Rottnest Island.

Habitat: Rocks and reefs or amongst kelp.

Distinctive features: Pectorla fin disk of this species is oval in shape and is about as wide as long. Anterior margins of the disk are gently convex and converge at a broad angle on the fleshy snout, tip of which is slightly protruding. Eyes are large and are followed by comma-shaped spiracles with rounded posterior margins. Mouth has 10 papillae on the floor and small teeth with roughly oval bases. Five pairs of gill slits are short. Pelvic fins are small and rounded. Tail ends in a short, deep, leaf-shaped caudal fin. A relatively large dorsal fin is positioned on the upper surface of the tail, followed shortly by the serrated stinging spine. Tail lacks lateral skin folds. Skin is entirely smooth. This species is slate-blue above with numerous whitish spots, blotches, and rings. Dorsal fin and margin of the caudal fin are brownish. Underside is plain white, becoming light brown on the tail. Largest known specimen is 60 cm long.

Food habits: Dietary composition has not yet been examined for this species (Kyne & White, 2006).

Reproduction: Little information is available on the natural history of this species. It is aplacental viviparous with the developing embryos sustained by maternally produced uterine milk. Litter sizes are probably small. Males attain sexual maturity at below 53 cm long.

Predators: Not reported.

Parasites: Not known.

IUCN conservation status: Least concern.

3.5.12   *UROLOPHUS CONCENTRICUS* (Osburn & Nichols, 1916)

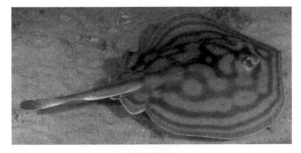

Common name: Bullseye round stingray.

Geographical distribution: Restricted distribution; only in the Gulf of California, Mexico.

Habitat: Rocky bottoms in coastal waters, bays, lagoons, and estuaries and also on sandy bottoms near reefs.

Distinctive features: This species has a flattened body with expanded pectoral fins which are fused with the body and head to form a round, flat disk. Tail is very short and is about equal to the length of the disk. Caudal fin is rounded and well developed. There are no dorsal fins. Its disk roughly circular in shape with a rounded snout and is generally light gray with blackish blotches and spots arranged in concentric rows. Two yellowish or cream bands are seen surrounding the disk. Skin is smooth, without spines, but a long, venomous spine is located approximately halfway down the length of the tail and is mainly used in defense. This species has a maximum size of 47.5 cm total length and 28 cm DW.

Food habits: Feeds on bottom-dwelling invertebrates, such as crustaceans, molluscs and worms, and small fish.

Reproduction: Like other stingrays, it is likely to be ovoviviparous, meaning the eggs hatch inside the female and the young are born live. Female gives birth to three to six young after a gestation period of around 3 months.

Predators: Not reported.

Parasites: Not known.

IUCN conservation status: Data deficient.

3.5.13 *UROLOPHUS CRUCIATUS* (Lacepède, 1804)

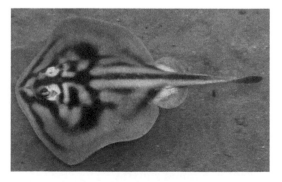

Common name: Crossback stingaree.

Geographical distribution: Temperate; eastern Indian Ocean: southern Australia.

Habitat: Marine; brackish; demersal; inshore waters, preferably muddy bottoms in the mouths of estuaries and in bays; rocky reefs.

Distinctive features: Pectoral fin disk of this species is slightly wider than long and is oval in shape with the anterior margins nearly straight and converging at a very obtuse angle. Snout is fleshy and blunt and is not protruding from the disk. Eyes are small and are immediately followed by teardrop-shaped spiracles, which have rounded to angular posterior rims. Mouth is small and strongly arched, containing 3–6 papillae on the floor and an additional patch of papillae is present on the outside of the lower jaw. Both upper and lower teeth are small with roughly oval bases and are arranged in a quincunx pattern. Five pairs of gill slits are short. Pelvic fins are small with rounded margins. It has rather a short tail which is flattened and oval in cross-section without lateral skin folds. A single serrated, stinging spine is positioned on top, about halfway along the tail's length. At the end of the tail is a very short, deep, leaf-shaped caudal fin. Skin is entirely devoid of dermal denticles. Body coloration is grayish to yellowish brown above with a pattern of dark markings, including a stripe which is running along the midline and crossed by three transverse bars. Underside is off-white, sometimes becoming slightly darker at the disk margin. Caudal fin is more grayish than the body and there may be dusky blotches on the tail. This species attains a maximum known total length of 50 cm.

Food habits: Feeds on crustaceans, molluscs, and worms.

Reproduction: Like other stingrays, this species is aplacental viviparous. When the developing embryos exhaust their supply of yolk, their mother provides them with nutrient-rich uterine milk through specialized extensions of the uterine epithelium called trophonemata. Females produce litters of 1–4 pups every other year. Birth length is between 10 and 15 cm and maturation length is between 20 and 32 cm. Both sexes mature at around 6 years of age and can live to at least 11 years.

Predators: Not reported.

Parasites

Monogenea: *Calicotyle urolophi*

Cestoda: *Acanthobothrium clarkeae*

Dolifusiella martini

Eutetrarhynchus martini

IUCN conservation status: Least concern.

3.5.14   *UROLOPHUS DEFORGESI,* Séret & Last, 2003

Common name: Chesterfield Island stingaree or Deforge's stingaree.

Geographical distribution: Western Central Pacific: Chesterfield Islands.

Habitat: Marine; bathydemersal; deep waters; continental slope.

Distinctive features: Pectoral fin disk of this species is diamond-shaped and 109–122% as wide as long, with broadly rounded outer corners and gently convex leading margins. Snout is fleshy and forms an obtuse angle with the tip slightly protruding. Eyes are medium-sized and somewhat closely spaced and are immediately followed by teardrop-shaped spiracles. Mouth is modestly sized and contains 7–8 papillae on the floor. There is also a patch of small papillae on the lower jaw. There are 28–33 upper tooth rows and 27–31 lower tooth rows. Five pairs of gill slits are short. Pelvic fins are small and rounded. Males have slender, pointed claspers.

Tail is somewhat flattened and measures 77–84% as long as the disk. There is a single dorsally placed, serrated stinging spine around halfway along the tail. There is no dorsal fin or lateral skin folds. Tail terminates in a long, low, leaf-shaped caudal fin. Skin is entirely devoid of dermal denticles. This species is plain yellowish-brown above with a dark caudal fin margin. Underside is white to cream and is darkening slightly at the fin margins. Largest known specimen has a maximum total length of 34 cm.

Food habits: Biology is largely unknown (Last & Marshall, 2006d).

Reproduction: Little is known of the natural history of this species. It is presumably aplacental viviparous with a small litter size, as in other members of its family. It is likely to have low fecundity (one to two young/ year). Newborns measure around 13 cm long and males reach sexual maturity at 29 cm long.

Predators: Not reported.

Parasites: Not known.

IUCN conservation status: Least concern.

3.5.15   *UROLOPHUS EXPANSUS,* McCulloch, 1916

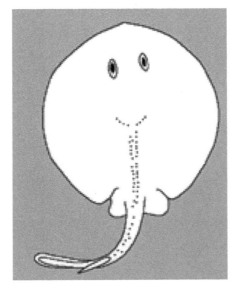

Common name: Wide stingaree.

Geographical distribution: Temperate; eastern Indian Ocean: endemic to Australia.

Habitat: Marine; bathydemersal; outer continental shelf and upper slope.

Distinctive features: This species has a diamond-shaped pectoral fin disk which is much wider than long with broadly rounded outer corners and trailing margins. Anterior margins are gently sinuous and converge at an obtuse angle on the fleshy, slightly protruding snout. Eyes are large and are immediately followed by comma-shaped spiracles with rounded posterior margins. Mouth is of medium size and contains 6–9 papillae on the floor. A narrow patch of papillae is also found on the lower jaw. Teeth are small with roughly oval bases. Five pairs of gill slits are short. Pelvic fins are small with rounded edges. Tail measures 71–93% as long as the disk and has a highly flattened base and well-developed lateral skin folds. There is no dorsal fin. Caudal fin is lance like and elongated. A serrated stinging spine is placed atop the tail about halfway along its length. Skin is completely smooth. Coloration of this species is dusky green above with two faint, bluish transverse lines behind the eyes and a pair of similar lines running obliquely outward from the eyes. Underside is white to yellow. Largest known specimen has a maximum total length of 52 cm.

Food habits: Little is known of the species' biology (Treloar, 2006).

Reproduction: Like other stingrays, it is aplacental viviparous, with the developing embryos sustained by uterine milk. Litter size is probably small as in related species. Males mature sexually at a DW of 30–36 cm (12–14 in.) and 7 years of age and have a maximum lifespan of 11 years. Females mature at a DW of about 40 cm.

Predators: Not reported.

Parasites

Cestoda: *Acanthobothrium adlardi*

Acanthobothrium clarkeae

Acanthobothrium odonoghuei

Acanthobothrium robertsoni

Prochristianella clarkeae

IUCN conservation status: Least concern.

3.5.16 *UROLOPHUS FLAVOMOSAICUS,* Last & Gomon, 1987

Common name: Patchwork stingaree.

Geographical distribution: Tropical; Indo-West Pacific: Australia.

Habitat: Marine; demersal; continental shelf and uppermost slope.

Distinctive features: This species has a diamond-shaped pectoral fin disk which is much wider than long with broadly rounded outer corners and nearly straight anterior margins that converge at an obtuse angle. Tip of the snout protrudes slightly past the disk. Small eyes are followed by comma-shaped spiracles with angular posterior rims. There are 8–14 stubby papillae on the floor of the large mouth, as well as a narrow patch of large papillae on the lower jaw. Teeth are small with roughly oval bases and there are five pairs of gill slits which are short. Pelvic fins are small with curved margins. Tail is short and flattened and it terminates in a short, deep, leaf-shaped caudal fin. A lateral skin fold runs along each side of the tail. Upper surface of the tail bears a rather large dorsal fin followed by a serrated stinging spine. Skin entirely lacks dermal denticles. Dorsal coloration of this species is consisting of a yellowish background with numerous large, dark brown rings. These rings become smaller and are less well defined toward the margins of the disk. Dorsal and caudal fins are light in adults and dark with nearly black margins in juveniles. This species can grow up to only 59 cm in total length.

Food habits: Biology of this species is largely unknown.

Reproduction: Virtually nothing is known of the natural history of the patchwork stingaree. Reproduction is presumably aplacental viviparous like in other stingrays, with the developing embryos sustained by maternally produced uterine milk. It is likely to have low fecundity (1–2 young/year) as with other urolophid species. Males mature sexually at 38 cm long (Last & Marshall, 2006e).

Predators: Not reported.

Parasites: Not known.

IUCN conservation status: Least concern.

3.5.17   *UROLOPHUS GIGAS,* Scott, 1954

Common name: Spotted stingaree.

Geographical distribution: Temperate; eastern Indian Ocean: endemic to Australia.

Habitat: Marine; demersal; continental shelf and inshore waters; estuaries; found partly covered with sand near seagrass beds.

Distinctive features: Pectoral fin disk of this species is oval in shape and is slightly longer than wide. Snout is fleshy and is usually smoothly rounded without a protruding tip. Eyes are tiny and are followed by much larger comma-shaped spiracles, which have rounded posterior rims. Floor of the small mouth bears 9–12 papillae. A further narrow patch of papillae is found on the lower jaw. Teeth are small with roughly oval bases. Five pairs of gill slits are short and pelvic fins are small and rounded. Tail is fairly thick and without lateral skin folds. A serrated stinging spine is placed atop the tail and is preceded by a prominent dorsal fin. Caudal fin is lance like, short, and deep. Skin is completely smooth. Disk is dark brown to black above and is becoming lighter toward the margins. There are 2–3 rows of small whitish spots that border the disk margin and may extend onto the tail. Much larger pale spots are also seen arranged in groups over the middle of the back. Tail has light spots in some adults. Dorsal and caudal fins are dark brown to black with whitish edges. Underside is nearly white. Most individuals have dusky blotches and wide bands bordering the lateral disk margins. This species can grow up to 80 cm total length.

Food habits: It feeds mainly on crustaceans.

Reproduction: As in other stingrays, it is presumably aplacental vivipa-rous with the developing embryos being nourished via maternal uterine milk. Females can produce litters of up to 13 pups. Males and females reach sexual maturity at 42 and 46 cm long, respectively.

Predators: Not reported.

Parasites: Unknown.

IUCN conservation status: Least concern.

3.5.18   *UROLOPHUS HALLERI,* Cooper, 1863 = *Urobatis halleri*

Common name: Haller's round ray.

Geographical distribution: Subtropical; eastern Pacific: Eureka in northern California, USA to Panama.

Habitat: Marine; demersal; sand or mud bottom off beaches and in bays and sloughs; around rocky reefs.

Distinctive features: This species has a nearly circular disk-shaped body with a tail which is shorter than the length of the disk. Snout termi-nates in a rounded point. Prominent pectoral fins (wings) are rounded. Dorsal fins are absent. However, a rounded caudal fin is present. A long spine is located approximately halfway down the length of the tail. This species is largely identified by its true tail fin. It is grayish brown, mottled, or spotted with dark blotches on the dorsal surface, fading to a pale yellow, orange, or white underside. Maximum total length and weight are 58 cm and 1.4 kg, respectively. Average DW is 8–25 cm.

Food habits: Feeds primarily on benthic invertebrates such as stomato-pods, amphipods, shrimp, and portunid crabs, and to a smaller extent, on polychaete worms and small fishes.

Reproduction: Development of this species is ovoviviparous, resulting in live birth. After a gestation period of approximately 3 months, a litter of 3–6 young are born in shallow waters. Each newborn measures 6.3–8.0 cm DW. Sexual maturity is reached between 2.6 and 3 years of age which relates to a DW of 14.6 cm. Life expectancy is believed to be about 8 years (Babel, 1967).

Predators: Predators of this species include the northern elephant seal (*Mirounga angustirostris*) and the black sea bass (*Stereolepis gigas*). Other potential predators of this ray include large carnivorous fishes including sharks.

Parasites: Not known.

IUCN conservation status: Least concern.

3.5.19 *UROLOPHUS JAMAICENSIS* (Cuvier, 1816) = *Urobatis jamaicensis*

Common name: Yellow stingray.

Geographical distribution: Western Atlantic from North Carolina to Florida and throughout the Gulf of Mexico off the Florida coast; Bahamas and Caribbean Sea.

Habitat: Shallow water in sandy or muddy habitats.

Distinctive features: This species is characterized by a round body. It has a well-developed caudal fin that extends around the tip of its tail. Tail spine is located just anterior to the caudal fin. Posterior edge of the pelvic fins is rounded. It lacks a dorsal fin. While individuals vary widely in color and pattern, dorsal side of the disk typically displays one of the following color schemes: a reticulate pattern of dark greenish or brown on a pale background, or equally a close set pattern of minute white, yellow, or golden spots on a dark green or brown background. Ventral side is yellowish or brownish-white. It grows to the size of 66 cm in length with a maximum DW of 35 cm (FLMNH Ichthyology Dept., https://www.flmnh.ufl.edu/fish/discover/species-profiles/urobatis-jamaicensis/).

Food habits: Feeds on worms, crabs or small fish.

Reproduction: This species is ovoviviparous like other stingrays. Gestation period is not currently known for this ray. A litter size of three to four has been reported. Males are sexually matured at a DW of 15–16 cm.

Predators: Any large carnivorous fish, especially sharks such as the tiger shark, is a potential predator of this species.

Parasites

Cestoda: *Acanthobothrium cartagenensis*

Phyllobothrium kingae

Rhinebothrium magniphallum

Discobothrium caribbensis

IUCN conservation status: Least concern.

3.5.20 *UROLOPHUS JAVANICUS* (Martens, 1864)

Common name: Java stingaree.

Geographical distribution: Tropical; eastern Indian Ocean: Java, Indonesia.

Habitat: Marine; demersal.

Distinctive features: This species has an oval pectoral fin disk which is slightly longer than wide. Leading margins of the disk are gently convex and converge at a blunt angle on the snout. Eyes are followed by larger, comma-shaped spiracles. Mouth is bow shaped and contains three papillae on the floor. Teeth are closely arranged with a quincunx pattern. Each tooth is small with a transverse ridge on the crown. Five pairs of gill slits are short. Pelvic fins are almost equal, with rounded corners. Tail is shorter than the disk and bears a prominent dorsal fin about halfway along its length. Immediately posterior to the dorsal fin, there is a serrated stinging spine. Tail ends in a leaf-shaped caudal fin. Skin is devoid of dermal denticles, though there are tiny white bumps on the upper central portion of the disk. This species is dark brown above, with many indistinct darker and lighter spots and pale below. It measures 33 cm long.

Food habits: Biology of this species is unknown.

Reproduction: Very little is known of the natural history of this species. It is presumably aplacental viviparous and is likely to have low fecundity (1–2 young/year) as with other urolophid species (Last & Marshall, 2006b).

Predators: Not reported.

Parasites: Not known.

IUCN conservation status: Critically endangered.

3.5.21 *UROLOPHUS KAIANUS,* Günther, 1880

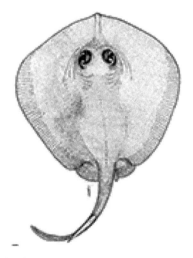

Common name: Kai stingaree.

Geographical distribution: Western Central Pacific: Kei Islands, Moluccas, Indonesia.

Habitat: Marine; bathydemersal; deep-waters; prefers muddy substrates.

Distinctive features: Pectoral fin disk of this species is diamond shaped and is much wider than long with narrowly rounded outer corners. Leading margins of the disk converge on the snout which is rounded and non-protruding. Eyes are immediately followed by teardrop-shaped spiracles. Teeth are small with roughly oval bases. Five pairs of gill slits are short and pelvic fins are small with rounded margins. Tail bears a skin fold running along either side. A leaf-shaped caudal fin is present at the end of the tail. A serrated stinging spine is located on top, about halfway along the tail's length. There is no dorsal fin. Skin is completely smooth. This species is plain yellowish-brown above, transitioning sharply to blackish on the upper surface of each eyeball. Maximum total length of this species is only 23 cm.

Food habits: Biology of this species is unknown.

Reproduction: Virtually nothing is known of the natural history of this species. It is presumably aplacental viviparous and it is likely to have low fecundity (1–2 young/year) as with other urolophid species.

Predators: Not reported.

Parasites: Not known.

IUCN conservation status: Data deficient.

3.5.22   *UROLOPHUS KAPALENSIS,* Yearsley & Last, 2006

Common name: Kapala stingaree.

Geographical distribution: Subtropical; southwest Pacific: New South Wales, Australia.

Habitat: Marine; benthopelagic; prefers rocky reefs and adjacent sandy flats, as well as seagrass beds.

Distinctive features: Pectoral fin disk of this species is more or less diamond shaped and is slightly wider than long. Leading margins of the disk are nearly straight and converge at an obtuse angle on the fleshy, barely protruding snout. Eyes are of modest size and are followed by teardrop-shaped spiracles with rounded posterior rims. Mouth is small and contains 5–7 papillae arranged in a W-shaped pattern on the floor. Lower jaw also bears a patch of prominent papillae arranged to form a series of transverse ridges. Teeth have rhomboid bases and are arranged in a quincunx pattern. There are 25 upper and 31–32 lower tooth rows. Five pairs of gill slits are S-shaped. Pelvic fins are small with rounded rear margins. Males have short, stout claspers. Slender, flattened tail terminates in a low, leaf-shaped caudal fin. There is a prominent skin fold running along each side. A very thin, serrated stinging spine is positioned atop the tail which is about halfway along its length. Immediate in front of the tail there is a long, low dorsal fin. Skin entirely lacks dermal denticles. It is greenish above and is becoming pinkish toward disk margins and bears a variable pattern of dark markings. Underside is off-white with a wide, dusky band around the disk margin. Tail is pale with a dark midline stripe above. Dorsal fin is greenish and the caudal fin is light with a dark edge in adults. Largest specimen measures 51 cm in total length (Yearsley & Last, 2006).

Food habits: Feeds primarily on benthic shrimp (mainly palaemonids) and amphipods (mainly ampeliscids). Important secondary prey groups are penaeid prawns and small bony fishes, while crabs, polychaete worms, and isopods are rarely consumed.

Reproduction: Like other stingrays, this species is aplacental viviparous with the developing embryos sustained by uterine milk produced by the mother. A litter consists of only a single pup, born at around 15 cm long. Males reach sexual maturity at 28–31 cm long.

Predators: Not reported.

Parasites: Unknown.

IUCN conservation status: Near threatened.

3.5.23   *UROLOPHUS LOBATUS,* McKay, 1966

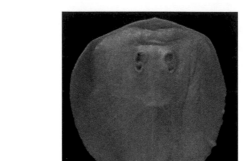

Common name: Lobed stingaree.

Geographical distribution: Temperate; eastern Indian Ocean: endemic to Australia.

Habitat: Marine; demersal; continental shelf; favors sandy flats and seagrass beds.

Distinctive features: It has a rounded pectoral fin disk which is much wider than long, with nearly straight leading margins. Snout is fleshy and forms an obtuse angle and its tip may protrude slightly past the disk. Medium-sized eyes are followed by teardrop-shaped spiracles which have a rounded posterior margin. Mouth is small and contains 9–10 papillae on the floor. A handful of papillae are also seen on the lower jaw. Small teeth have roughly oval bases. Five pairs of gill slits are short. Pelvic fins are small and rounded. Tail is slender and is very flattened with a prominent horizontal skin fold on either side. There is a dorsally positioned, serrated stinging spine near the caudal fin, which is long, narrow, and leaf shaped. There is no dorsal fin. Skin entirely lacks dermal denticles. This species is yellowish brown above and is becoming slightly lighter at the lateral margins of the side, and white below. Caudal fin becomes dark toward the tip. Males and females have a DW of 24 and 27 cm, respectively. It grows to a maximum size of 38 cm total length.

Food habits: Crustaceans, in particular mysids, amphipods, shrimps, and cumaceans form the main food items. This species also feeds on polychaete worms, small bony fishes, and rarely molluscs.

Reproduction: Like other stingrays, this species is aplacental viviparous. Females have a single functional uterus on the left and an annual reproductive cycle. Gestation period lasts 10 months. Usually only a single pup (rarely two) develops to term. Embryo is initially nourished by an external yolk sac and by 6 months of age, mother begins to deliver nutrient-rich uterine milk through thread-like extensions of the uterine epithelium called trophonemata. Newborns measure 11 cm across. Sexual maturity is reached by males at 16 cm across and 2 years of age, and by females at 20 cm across and 3 years of age. Maximum life-span is 12 and 14 years for males and females, respectively.

Predators: Not reported.

Parasites

Cestoda: *Acanthobothrium adlardi*

Acanthobothrium odonoghuei

Acanthobothrium robertsoni

Acanthobothrium rohdei

IUCN conservation status: Least concern.

3.5.24   UROLOPHUS MACULATUS (Garman, 1913)

Common name: Spotted round ray.

Geographical distribution: Subtropical; eastern central Pacific: southern Baja California, Mexico, and Gulf of California.

Habitat: Marine: demersal; shallow sand and mud bottoms; sea grass beds; near rocky reefs and in bays.

Distinctive features: This species has a flattened body with expanded pectoral fins which are fused to the head and body to form a round, flat disk. It has a short tail which is equal to or slightly less than the length of the disk and a well-developed, rounded caudal fin. No dorsal fins are present. Disk of this species is roughly circular in shape with a slightly angular snout. Skin is smooth and upper surface of the body is brown or brownish-gray in color with variable, irregular, relatively widely spaced dark blotches and spots. A long, pine is located halfway along the upper side of the tail. Maximum size of this species is 42 cm total length and 26 cm DW.

Food habits: Diet typically includes bottom-dwelling invertebrates, such as crustaceans, molluscs, and worms, as well as small fish.

Reproduction: Like other stingrays, it is likely to be ovoviviparous and female is retaining the eggs inside the body until they hatch, and then giving birth to live young. Each female gives birth to 3–6 young after a gestation period of around 3 months.

Predators: Not reported.

Parasites: Not known.

IUCN conservation status: Data deficient.

3.5.25   UROLOPHUS MITOSIS, Last & Gomon, 1987

Image not available

Common name: Mitotic stingaree.

Geographical distribution: Tropical; Eastern Indian Ocean: known only from off Port Hedland, northwestern Australia.

Habitat: Marine; bathydemersal; fine sediment at the edge of the continental shelf.

Distinctive features: Pectoral fin disk of this species is more or less diamond shaped and is slightly wider than long. Leading margins are nearly straight while the outer corners and trailing margins are rounded. Snout forms an obtuse angle with the tip extending slightly past the disk. Eyes are large and are followed by spiracles which are comma shaped with rounded or angular posterior rims. Mouth is somewhat large with 3–4 papillae on the floor. A few papillae are also seen on the lower jaw.

Teeth are small with roughly oval bases and five pairs of gill slits are short. Pelvic fins are small and rounded. Slender tail is strongly flattened and bears a subtle skin fold along each side. A serrated stinging spine is positioned atop the tail about halfway along its length. Caudal fin is elongated and lance like. Skin is entirely smooth. Upper surface of the disk is light green which is becoming reddish toward the margins. Several large, light blue blotches are seen on the disk. Underside and caudal fin are uniformly light. Largest specimen measures 29 cm in total length.

Food habits: Feeds on crustaceans, molluscs and worms, as well as small fish.

Reproduction: Virtually nothing is known of the natural history of this species. It is presumably aplacental viviparous, with the developing embryos sustained by uterine milk produced by the mother, like other stingrays. Litter size is probably small (1–2 young/year), as in related species. Males reach sexual maturity at a total length of 25 cm.

Predators: Not reported.

Parasites: Not reported.

IUCN conservation status: Least concern.

3.5.26 *UROLOPHUS NEOCALEDONIENSIS,* Séret & Last, 2003

Common name: New Caledonian stingaree.

Geographical distribution: Western Central Pacific: New Caledonia (including the Chesterfield Islands) and northern part of the Norfolk Ridge).

Habitat: Marine; bathydemersal; deep waters; continental slope.

Distinctive features: This species has a rhomboid pectoral fin disk which is as wide as long with broadly rounded outer corners and strongly convex anterior margins. Snout is fleshy and broad with a protruding tip. Eyes are of medium size and are immediately followed by comma-shaped spiracles. Mouth is modestly sized and bears 7–10 papillae on the floor. These papillae are arranged in a "W" pattern. There is also a patch of small papillae on the outside of lower jaw. Teeth are arranged in 27–34 rows in the upper jaw and 24–31 rows in the lower jaw. Five pairs of gill slits are short. Pelvic fins are small with rounded margins. Males have very thick claspers with rounded tips. Tail is fairly long. The serrated stinging spine is placed on the upper surface of the tail about halfway along its length. There is no dorsal fin but there may be a subtle fold of skin running along either side of the tail. Terminal leaf-shaped caudal fin is short and deep. Skin is completely smooth. This species is uniformly grayish-brown to olive brown above and underside is whitish with a broad dusky band along the lateral and posterior disk margins. Margins of pelvic, dorsal, and caudal fins and the tip of the snout are dark. Largest specimen measures 37 cm in total length (Séret & Last, 2003).

Food habits: Biology of this species is unknown (Last & Marshall, 2006h).

Reproduction: Little is known of the New Caledonian stingaree's natural history. It is presumed to be aplacental viviparous, bearing small litters, like other members of its family. Newborns measure roughly 13 cm long. Males attain sexual maturity at about 30 cm long.

Predators: Not reported.

Parasites: Not known.

IUCN conservation status: Least concern.

3.5.27 *UROLOPHUS ORARIUS*, Last & Gomon, 1987

Common name: Coastal stingaree.

Geographical distribution: Subtropical; eastern Indian Ocean: eastern Great Australian Bight.

Habitat: Marine; demersal shallow coastal waters.

Distinctive features: Body of this species is flattened and disk shaped. Pectoral fins are broadly expanded and are fused with the head and trunk. It possesses a remarkably circular-shaped body disk. It has a long, relatively narrow tail which is distinctly demarcated from the disk-like body. It possesses one or more long spines which are located about half way down their tail. This species is grayish-brown with dark mottling on its upper surface and paler underneath. This species has a maximum total length of 31 cm only.

Food habits: It feeds on bottom-dwelling fishes, worms, shrimp, and molluscs.

Reproduction: Like other species of stingrays, this species is also ovoviviparous. It gives birth to between two and four live young per litter. It has a gestation period of around 3 months.

Predators: Not reported.

Parasites: Not known.

IUCN conservation status: Endangered.

3.5.28   *UROLOPHUS PAPILIO,* Séret & Last, 2003

Common name: Butterfly stingaree.

Geographical distribution: Western Central Pacific: Chesterfield Islands.

Habitat: Marine; bathydemersal; deep waters; continental slope.

Distinctive features: This species has a diamond-shaped pectoral fin disk which is as wide as long, with broadly rounded outer corners. Snout forms an obtuse angle and has a protruding tip. Eyes are modest in size and are immediately followed by teardrop-shaped spiracles. Mouth is moderately large and contains 10–13 papillae in a row across the floor. There are 24–28 upper and 26–31 lower tooth rows. Five pairs of gill slits are short. Pelvic fins are small and rounded. Males have slightly pointed claspers. Tail is flattened at the base and tapers rapidly. There is a serrated stinging spine placed atop the tail and is about midway along its length, which is preceded by a low dorsal fin. Tail may also have slight ridge of skin running along each side and terminates in a very short, deep, leaf-shaped caudal fin. Skin entirely lacks dermal denticles. This species is yellowish to brownish above and underside is white to cream with a wide darker band along the lateral and posterior disk margins. Dorsal and caudal fins are dusky edged. It grows to 40 cm in total length (Séret & Last, 2003).

Food habits: Biology of this species is unknown.

Reproduction: Little is known of the natural history of this species. It is presumably aplacental viviparous with a small litter size, like other stingarees. Young are born at about 14 cm long. Males mature sexually at about 31 cm long (Last & Marshall, 2006f).

Parasites: Not known.

IUCN conservation status: Least concern.

3.5.29   *UROLOPHUS PAUCIMACULATUS,* Dixon, 1969 = *Trygonoptera paucimaculatus*

Common name: Sparsely spotted stingaree.

Geographical distribution: Temperate; eastern Indian Ocean: endemic to Australia.

Habitat: Marine; demersal; Variety of sandy or seagrass-bottomed habitats, ranging from very shallow, sheltered bays and inlets to the open continental shelf.

Distinctive features: This species has a more or less diamond-shaped pectoral fin disk which is wider than long with rounded outer corners. Anterior margins of the disk are nearly straight and converge at an obtuse angle on the fleshy snout. Small eyes are immediately followed by comma-shaped spiracles with angular or rounded posterior rims. Small mouth contains five or six papillae on the floor and most of them have forked tips. Additional small papillae are also present on the outside of the lower jaw. Teeth in both jaws are small with roughly oval bases and are arranged in a quincunx pattern. Five pairs of gill slits are short. Pelvic fins are small and rounded. Tail is very flattened at the base and slender toward the tip, which bears a deep, leaf-shaped caudal fin. There is a prominent fold of skin running along either side of the tail and a serrated stinging spine is located on the upper surface about halfway along its length. There is no dorsal fin. Skin is completely devoid of dermal denticles. This species is uniformly light gray above with a darker V-shaped marking between eyes and white below with slightly darker lateral disk margins. Largest individual measures 57 cm total length.

Food habits: Adults feed on worms, crabs, shrimp, and small bony fishes while juveniles take small crustaceans.

Reproduction: Like other stingrays, it is aplacental viviparous. Once the developing embryos exhaust their supply of yolk, mother supplies them with nutrient-rich histotroph (uterine milk) via specialized extensions of the uterine epithelium called trophonemata. Females have a single functional ovary and uterus on the right side and an annual reproductive cycle. Gestation period lasts approximately 1 year, and the newborns measure about 15–16 cm long. Males reach sexual maturity at around 28 cm long and two and a half years of age and females at around 27 cm long and 3 years of age. The maximum lifespan is at least 8 years for males and 9 years for females.

Predators: Not reported.

Parasites

Monogenea: *Calicotyle urolophi*

Cestoda: *Acanthobothrium adlardi*

Acanthobothrium clarkeae

Acanthobothrium urolophi

Dollfusiella geraschmidti

Trimacracanthus aetobatidis

IUCN conservation status: Least concern.

3.5.30   UROLOPHUS PIPERATUS, Séret & Last, 2003

Common name: Coral sea stingaree.

Geographical distribution: Western Central Pacific: northern Queensland, Australia.

Habitat: Marine; bathydemersal; deep-waters; continental shelf and upper continental slope.

Distinctive features: This species has a diamond-shaped pectoral fin disk which is wider than long, with nearly straight leading margins and rounded outer corners. Tip of the fleshy, triangular snout protrudes past the disk. Eyes are rather large and are followed by teardrop-shaped spiracles with rounded to angular posterior rims. Mouth is of medium size and contains 7–9 papillae on the floor. Lower jaw also bears a patch of minute papillae. Teeth number 32–35 rows in the upper jaw and 30–39 in the lower jaw and are small with roughly rhomboid bases. Five pairs of gill slits are short. Pelvic fins are small and rounded posterior. Males have blunt claspers. Tail is flattened at the base. A serrated stinging spine is located atop the tail about halfway along its length with a low dorsal fin just in front. Tail ends in a short, deep, leaf-shaped caudal fin. Skin is devoid of dermal denticles. Individuals are generally light gray or brown above with tiny dark dots. Dorsal and caudal fins are brown with blackish margins. Underside is white with dusky lateral disk margins and/or a few dark blotches on the tail. Largest specimen is 48 cm in total length (Séret & Last, 2003).

Food habits: It feeds primarily on benthic invertebrates.

Reproduction: Virtually nothing is known of the natural history of this species. It is presumably aplacental viviparous like other stingrays with developing embryos provided with maternally produced histotroph (uterine milk). Newborns measure 12 cm. Litter size is probably small based on related species. Males and females attain sexual maturity at 23 and 27 cm long, respectively.

Predators: Not reported.

Parasites: Not known.

IUCN conservation status: Least concern.

3.5.31 *UROLOPHUS SUFFLAVUS,* Whitley, 1929

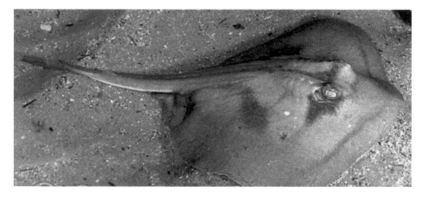

Common name: Yellowback stingaree.

Geographical distribution: Temperate; southwest Pacific: Australia, from Queensland to New South Wales.

Habitat: Marine; demersal; soft substrates on the continental shelf and upper slope.

Distinctive features: This species has a flattened pectoral fin disk which is about as wide as long with rounded corners and straight anterior margins. There is a skirt-like nasal curtain in front of the mouth without lateral lobes. Tail is short and stout with a serrated spine. Tail ends in a small caudal fin. There are no dorsal fins or lateral folds. Skin is devoid of dermal denticles. Coloration is uniformly yellowish above. This species attains a maximum total length of 42 cm.

Food habits: Feeds on bottom-dwelling fishes, worms, shrimp and other small organisms in the substrate around them, with some species able to eat hard-shelled molluscs and crustaceans.

Reproduction: Very little is known about the biology of this species but it is ovoviviparous and female usually gives birth to between two and four live young per litter. Gestation period is around 3 months. Males mature at a length of 23 cm.

Predators: Not reported.

Parasites: Not known.

IUCN conservation status: Vulnerable.

3.5.32   *UROLOPHUS VIRIDIS,* McCulloch, 1916

Common name: Greenback stingaree.

Geographical distribution: Temperate; Indo-West Pacific: Australia (from southern tip of Queensland southward to Tasmani).

Habitat: Sea; demersal; freshwater; upper continental slope; outer continental shelf.

Distinctive features: This species has a diamond-shaped pectoral fin disk which is wider than long with broadly rounded outer corners. Leading margins of the disk are nearly straight and converge at an obtuse angle on the fleshy snout. Tip of the snout protrudes slightly past the disk. Eyes are large and followed by comma-shaped spiracles with rounded posterior margins. Medium-sized mouth contains 4–7 variably shaped papillae on the floor. Additional papillae are seen in a narrow strip on the lower jaw. Teeth are small with roughly oval base, and five pairs of gill slits are short. Pelvic fins are small and rounded on their trailing margins. Tail is flattened at the base. A prominent skin fold runs along each side of the tail. A deep, lance-shaped caudal fin is found at the end. Upper surface of the tail bears a serrated stinging spine about halfway along its length. There is no dorsal fin. Skin entirely lacks dermal denticles. This species is plain light green above and coloration is becoming lighter toward the edge of the disk. Underside is white and purplish or pinkish toward the lateral margins of the disk. Ventral and lateral disk margins may also have a dark brown edge or blotches. Caudal fin is olive colored. The maximum recorded total length of this species is 51 cm.

Food habits: Preys mainly on polychaete worms and crustaceans.

Reproduction: Like other stingrays, it is aplacental viviparous with the developing embryos sustained via histotroph (uterine milk) produced by the mother. Females produce 1–3 pups per year, following a gestation period which is lasting 10–12 months. Males and females attain sexual maturity at 28 cm and 26–31 cm long, respectively (Last & Marshall, 2006g).

Predators: Not reported.

Parasites: Monogenea: *Calicotyle* sp.

IUCN conservation status: Vulnerable.

3.5.33   *UROLOPHUS WESTRALIENSIS,* Last & Gomon, 1987

Image not available

Common name: Brown stingaree.

Geographical distribution: Eastern Indian Ocean: Western Australia.

Habitat: Marine; demersal; deep-water; outer continental shelf.

Distinctive features: Pectoral disk of this species is diamond-shaped and is slightly wider than long with broadly rounded outer corners. Anterior margins are nearly straight and converge at an obtuse angle on the snout, which protrudes slightly from the disk. Eyes are of modest size and are followed by teardrop-shaped spiracles with rounded posterior rims. There are 5–6 small papillae on the floor of the fairly large mouth along with a few papillae on the lower jaw. Small teeth have roughly oval bases. Five pairs of gill slits are short and pelvic fins are small and rounded. Tail is rather short with a serrated stinging spine on top about halfway along its length. A short, deep, leaf-shaped caudal fin is seen at the end of the tail. Skin is completely smooth. Upper surface of body is light yellow or brown. There may be faint, darker bars running across the eyes, gill region, and middle of the back. Underside is whitish and the caudal fin is yellow with a black margin. Largest known specimen measures 36 cm in total length.

Food habits: Preys mainly on polychaete worms and crustaceans.

Reproduction: Little is known of the natural history of this species. Reproduction is aplacental viviparous with the developing embryos sustained by histotroph (uterine milk) produced by the mother. Litter size is not known. Newborns measure 10–13 cm long. Males mature sexually at 24 cm long.

Predators: Not reported.

Parasites: Not reported.

IUCN conservation status: Least concern.

3.5.34   *UROTRYGON ASPIDURA* (Jordan & Gilbert, 1882)

Phylum: Chordata   Subphylum: Vertebrata
Class: Chondrichthyes   Subclass: Elasmobranchii
Order: Myliobtiformes   Family: Urolophidae

Common name: Panamic stingray.

Geographical distribution: Tropical; eastern Central Pacific: Costa Rica and Panama.

Habitat: Marine; demersal; coastal waters over sandy and muddy bottoms; rocky reef areas.

Distinctive features: This species has a flat, oval body which is larger in depth than in length. Disk has fairly straight front margins. Color varies from grayish brown in the center to reddish brown at the disk edge. Underside is off white with a pinkish tinge. Eyes are small and snout is pointed. Pelvic fins have straight rear edges. Tail is slightly longer than disk length and has a narrow rounded caudal fin. A row of six large thorns is running from the base to the middle of the tail which is followed by a large venomous spine. Disk is sparsely covered with small denticles. It reaches a maximum total length of 50 cm and a DW of 23 cm (Mexico Fish, Flora, & Fauna, http://www.mexfish.com/mexico/panamic-stingray/).

Food habits: Feeds on small crustaceans, small fish, mussels, and worms.

Reproduction: Reproduction is aplacental viviparous with the developing embryos sustained by histotroph (uterine milk) produced by the mother. Life history of this species is poorly known and no other biological data are available (Valenti & Robertson, 2009).

Predators: Not reported.

Parasites: Not known.

IUCN conservation status: Data deficient.

3.5.35   *UROTRYGON CAUDISPINOSUS,* Hildebrand, 1946

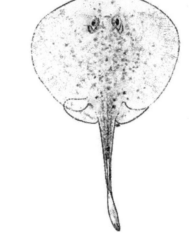

Common name: Spine-tailed roundray.

Geographical Distribution: Tropical; southeast Pacific: Peru.

Habitat: Marine; demersal.

Distinctive features: Pectoral fins of this species form a circular disk whose width is greater than the length. Pointed snout forms an obtuse angle and supports the edge of the disk. Behind the eyes are spiracles. Tail ends are with a leaf-shaped tail fin. In the middle of the caudal peduncle, there is a thorn. Pelvic fins are rounded. It grows to a maximum total length of 28.8 cm (Chirichigno & Vélez, 1998).

Biology of this species is unknown.

IUCN conservation status: Not evaluated.

3.5.36 *UROTRYGON CHILENSIS* (Günther, 1872)

Common name: Blotched stingray.

Geographical distribution: Subtropical; eastern Pacific: Gulf of California to Peru and Chile.

Habitat: Marine; demersal; soft bottoms of shallow water.

Distinctive features: Pectoral fins of rays form a circular disk whose width is greater than the length. Pointed snout forms an obtuse angle and supports the edge of the disk. Behind the eyes are spiracles. Pectoral fin disk is not more than 1.3 times as broad as long. Tail is slender and about as long as disk length. Dorsal fin is absent. Caudal fin is distinct. Tail ends are with a leaf-shaped tail fin. In the middle of the caudal peduncle, there is a poisonous thorn. Pelvic fins are rounded. It has a tan body with brown spots. Maximum recorded total length of this species 41.9 cm.

Food habits: Feeds mainly on small crustaceans, molluscs, polychaete worms, and fishes.

Reproduction: Reproduction is aplacental viviparous with the developing embryos sustained by histotroph (uterine milk) produced by the mother. No other information relating to this species is available (Lamilla, 2004).

Predators: Not reported.

Parasites

Cestoda: *Acanthobothrium campbelli*

Escherbothrium molinae

IUCN conservation status: Data deficient.

3.5.37 *UROTRYGON CIMAR,* López & Bussing, 1998

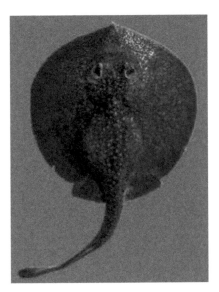

Common name: Denticled roundray.

Geographical distribution: Tropical; eastern Central Pacific: between Corinto, Nicaragua, and Gulf of Nicoya, Costa Rica.

Habitat: Marine; demersal; tide pools; shallow waters; soft bottoms.

Distinctive features: This species has a round disk. Tail is short. Dorsal surface of disk and tail is yellow-brown or tan and are irregularly spotted with brown or black blotches. Ventral surface of disk is white with lateral margins of disk. Posterior border of pelvic fins is dark brown or gray. Small denticles densely cover dorsal surface of disk and tail. Denticles are only slightly enlarged near midline. Pupillary operculum is a small triangle and its apex is not reaching ventral margin of pupil. This species grows to a maximum size of 38.2 cm in total length (López & Bussing, 1998).

Food habits: Diet of this species consists of mobile benthic worms; mobile benthic crustacea (shrimps/crabs); mobile benthic gastropods/bivalves; octopus/squid/cuttlefish; bony fishes.

Reproduction: Reproduction is aplacental viviparous with the developing embryos sustained by histotroph (uterine milk) produced by the mother. Biology of this species is largely unknown.

Predators: Not reported.

Parasites: Not reported.

IUCN conservation status: Not evaluated.

3.5.38   *UROTRYGON MICROPHTHALMUM,* Delsman, 1941

Common name: Small-eyed round stingray.

Geographical distribution: Tropical; western Atlantic: Venezuela and the mouth of the Amazon River.

Habitat: Marine; demersal.

Distinctive features: This species has a disk which is angular in front and naked. Snout is long and pointed. Tail is longer; pectorals are angular and caudal fin is narrower. Upper surface is gray and lower surface is whitish. This species is characterized by a smooth surface of the disk and devoid of spines and thorns on the back and belly. Its color ranges in olive green to dark brown shades on the back and belly is whitish. Central portion is light with a dark edge. Depressed tail has a serrated stinger on the lateral margins and caudal fin is found inserted in its terminal part. This species grows to a maximum size of 12.3 cm DW, total length of 23.2 cm and weight of 82 g (Maranhao fish; Uyeno et al., 1983).

Food habits: Feeds mainly on cumaceans and polychaetes.

Reproduction: Reproduction is aplacental viviparous with the developing embryos sustained by histotroph (uterine milk) produced by the mother. This species is also known for its abnormal hermaphroditism and is classified as a pseudo-hermaphrodite.

Predators: Not reported.

Parasites: Unknown.

IUCN conservation status: Least concern.

3.5.39   *UROTRYGON MUNDA*, Gill, 1863

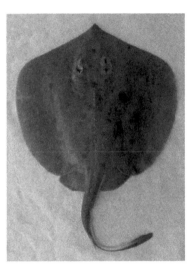

Common name: Munda round ray, spiny stingray.

Geographical distribution: Tropical; eastern Central Pacific: Central America.

Habitat: Marine; demersal; coastal sandy and muddy bottoms.

Distinctive features: These fish have rounded flat bodies with disks which are about equal in length and depth. They have a uniform light brown to yellowish brown color with tails having lighter edges. Undersides are off-white. Disks have front margins which are straight to slightly convex. Heads have a short weakly pointed snout. Small eyes and spiracles are on top of the heads and mouths, nostrils, and gill slits are on the ventral sides. Their slender tails are longer than disk length and have an elongated oval caudal fin. One or two rows of 18–32 recurved spines are seen along the mid-back that extend from mid-disk to tail spine. There is one large venomous spine on mid-tail. Disks and tails are densely covered with relatively large, strong, and recurved denticles with star-like bases. These spiny stingrays have a maximum total length of 40 cm with maximum DW of 24 cm (Mexico Fish, Flora, & Fauna, http://www.mexfish.com/mexico/spiny-stingray/).

Food habits: Feeds on small crustaceans, small fish, mussels, and worms.

Reproduction: Like other myliobatiformes, this species multiplies by ovoviviparity.

Predators: Not reported.

Parasites: Unknown.

IUCN conservation status: Data deficient.

3.5.40 *UROTRYGON NANA,* Miyake & McEachran, 1988

Common name: Dwarf round ray, pygmy round ray.

Geographical distribution: Tropical; eastern central Pacific: central and southern Mexico; Costa Rica.

Habitat: Marine; demersal; soft bottoms; close to mangrove forests.

Distinctive features: Disk of this species is oval and is relatively wide. Eyes are very small. Pectorals are continuous around head and tail is slender. Tail length is greater than disk length. Tail fin is elongated and oval. There is a large spine on tail and it lacks dorsal fin. Small curved denticles either uniformly cover entire disk or densely cover edges of snout and disk. Mid disk, however, is sparsely covered by denticles. Denticles are also seen on top and sides of tail and on upper lobe of tail fin. Midline is devoid of thorns. Upper surface is pale tan with indistinct fine meshwork of darker lines particularly in center of disk and tail base. Lower surface is white with broad dark gray-brown edging to disk. It grows to a maximum size of 25 cm total length and DW of 15 cm (Miyake & McEachran, 1988).

Food habits: Little else is known of its biology (Robertson et al., 2009).

Reproduction: These rays, which are oviparous like other stingrays, reach sexual maturity at 17 cm total length. Fecundity is very low with only one embryo per female. Size at birth is ~6 cm total length. Little else is known of its biology.

Predators: Not reported.

Parasites: Unknown.

IUCN conservation status: Data deficient.

3.5.41   UROTRYGN PERUANUS, Hildebrand, 1946

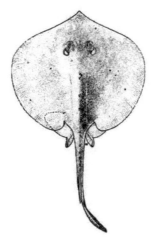

Common name: Peruvian stingray.

Geographical distribution: Tropical; southeast Pacific: Peru.

Habitat: Marine; demersal; soft bottom.

Distinctive features: Disk of this species is very much flattened, rounded, and a little wider than long. Its front edges are straight and oblique. Snout is pointed and is projecting a little. Eyes and spiracles are located on top of head. Mouth and gills are seen on the underside. Floor of the mouth is with fleshy papillae. Jaw teeth are small and numerous in bands. Pelvic fins are with broadly rounded rear edges. Tail is slender with a spine which has about snout length and large serrations particularly on tip half. Tail fin is elongated and oval in shape. Upper surface of disk is smooth except for a patch of prickles on snout. Coloration is gray above and whitish below. Dorsal disk has few scattered small dark spots. This species grows to a maximum total length of 27.6 cm.

Nothing is known of its biology.

IUCN conservation status: Not evaluated.

3.5.42   UROTRYGON RETICULATA, Miyake & McEachran, 1988

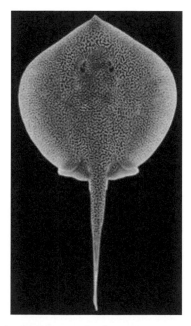

Common name: Reticulated round stingray.

Geographical distribution: Tropical; eastern Central Pacific: only from Panama.

Habitat: Marine; demersal; soft bottoms.

Distinctive features: Entire dorsal body of this species has fine tan to brownish vermiculations. Pattern is more diffuse consisting of speck-like markings, on extreme margin of disk, pelvic fins, and dorsal lobe of caudal fin. Tail has a spine. It grows to a maximum size of 24.1 cm total length (Miyake & McEachran, 1988).

Very little is known of its biology (Robertson & Valenti, 2009b).

IUCN conservation status: Vulnerable.

3.5.43   *UROTRYGON ROGERSI* (Jordan & Starks, 1895)

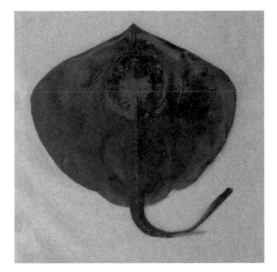

Common name: Roger's round ray, thorny stingray.

Geographical distribution: Tropical; eastern Central Pacific: southern Gulf of California to Ecuador.

Habitat: Marine; demersal; soft bottoms of coastal waters.

Distinctive features: This species has angular and somewhat diamond-shaped disks with straight front margins and a slightly projecting pointed snout. Coloration of body is uniform light brown to yellowish brown without distinctive markings. Undersides are off-white. Small eyes and spiracles are on top of head and mouth, nostrils, and gill slits are on the ventral sides. Slender tail is longer than half their total body length and its narrow pointed caudal fin has a rounded end. Dorsal side is covered with small denticles. There are also a row of approximately 30 thorns that run from the nape along the middle of the back and on top of the tail to the stinger. A large venomous spine is found on mid-tail. It attains a maximum total length of 46.2 cm.

Food habits: Feeds on crustaceans and small fishes.

Reproduction: This ovoviviparous species has a reproductive strategy based on low fecundity, a rapid reproductive cycle (short ovulation and gestation time), three birth peaks per year, and large embryos.

No other information is available on its biology or ecology of this species (Valenti, 2009c).

IUCN conservation status: Data deficient.

3.5.44   *UROTRYGON SIMULATRIX,* Miyake & McEachran, 1988

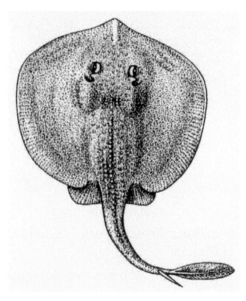

Common name: Fake round ray.

Geographical distribution: Tropical; eastern Central Pacific: only from Gulf of Panama.

Habitat: Marine; demersal; soft bottoms.

Distinctive features: Entire dorsal disk and tail of this species are covered with high stout cone-shaped and slightly recurved denticles. Denticles are found enlarged toward midline and are forming one or two continuous rows on midline of disk and tail. Dorsal surface of disk and tail is uniformly dark grayish brown and are relatively broadly edged with yellowish white. It grows to a maximum total length of 26.7 cm (McEachran, 1995; Miyake & McEachran, 1988).

Little is known about the life history parameters of this species (Robertson & Valenti, 2009a).

Parasites: Cestoda: *Parachristianella dimegacantha*

IUCN conservation status: Vulnerable.

3.5.45   UROTRYGON VENEZUELAE, Schultz, 1949

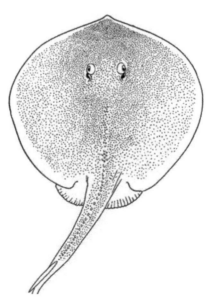

Common name: Venezuela round stingray.

Geographical distribution: Tropical; western central Atlantic: Gulf of Venezuela.

Habitat: Marine; demersal; inshore coastal species.

Distinctive features: This species has a prickly back with a median row of slightly larger spines from the nuchal region rearward along disk and anterior part of tail. Disk is wide and tail is long. Body is plain grayish above without markings and is cream colored below. It attains a maximum total length of 25.5 cm (Compagno, 1999).

Nothing else is known concerning its biology and ecology of this species (Kyne & Valenti, 2007).

Parasites

Cestoda: *Acanthobothrium urotrygoni*

Rhinebothrium magniphallum

IUCN conservation status: Near threatened.

3.6   SIXGILL STINGRAY *(HEXATRYGONIDAE)*

3.6.1   HEXATRYGON BICKELLI, Heemstra & Smith, 1980 = *Hexatrygon brevirostra, H. longirostrum, H. taiwanensis, H. yangi.*

Phylum: Chordata   Subphylum: Vertebrata
Class: Chondrichthyes   Subclass: Elasmobranchii
Order: Myliobtiformes   Family: Hexatrygonidae

Common name: Sixgill stingray.

Geographical distribution: Indo-Pacific: South Africa eastward to Japan, Australia and Hawaiian Islands.

Habitat: Marine; bathydemersal; deep-water; soft bottoms of upper continental slopes.

Distinctive features: This species has a bulky, flabby body with a rounded pectoral fin disk which is longer than wide. Triangular snout is much longer in adults and is filled with a clear gelatinous material. Tiny eyes are placed far apart and well ahead of the larger spiracles. Mouth is wide and almost straight. In either jaw are 44–102 rows of small, blunt teeth arranged in a quincunx pattern. Teeth are numerous in adults. Six pairs of small gill slits appear on the underside of the disk. Pelvic fins are rather large and rounded. Tail is moderately thick. One or two serrated stinging spines are present on its dorsal surface, well back from the base. Average STL is 64 mm for males and 124 mm for females and total number of serrations varies from 61 to 139 in these sexes. End of the tail bears a long, low leaf-shaped caudal fin which is nearly symmetrical above and below. Skin is delicate and dermal denticles are completely absent. Disk is purplish to pinkish brown above, darkening slightly at the fin margins. Belly is white with dark margins on the pectoral and pelvic fins. Snout is translucent, and tail and caudal fin are almost black. Maximum total length of this species is 1.4 m (Schwartz, 2007).

Food habits: Unknown.

Reproduction: Reproduction in this species is viviparous, with a litter size between two and five pups. Newly born rays measure 48 cm long. Both males and females mature sexually at 1.1 m long.

Predators: Not reported.

Parasites: Not reported.

IUCN conservation status: Least concern.

3.7   DEEPWATER STINGRAY (PLESIOBATIDAE)

3.7.1   PLESIOBATIS DAVIESI (Wallace, 1967) = *Urotrygon davies*

Phylum: Chordata   Subphylum: Vertebrata
Class: Chondrichthyes   Subclass: Elasmobranchii
Order: Myliobtiformes   Family: Plesiobatidae

Common name: Deep water stingray.

Geographical distribution: Indo-Pacific: South Africa, Mozambique, southern India, east to the Philippines, north to Japan, and south to Australia; Hawaiian Islands.

Habitat: Marine; bathydemersal; deep waters; outer shelf and upper slope; soft bottoms.

Distinctive features: This species has a flabby body with enlarged pectoral fins forming a disk which is usually longer than it is wide. Leading margins of the disk converge at a broad angle. Snout is thin and its tip protrudes slightly from the disk. Small eyes are positioned just ahead of spiracles, which have angular posterior rims. Wide, straight mouth contains 32–60 tooth rows in either jaw. Each tooth is small with a low, blunt cusp. In adult males, teeth at the center are sharp and backward-pointing. Five pairs of gill slits are small and placed beneath the disk. Pelvic fins are small and have blunt outer corners. Tail is thick. Lateral skin folds and dorsal fins are absent. One or two serrated stinging spines are present atop the tail. Average STL of males is 114 mm and 81 mm in females. Total number of serrations in these sexes is 114 and 81, respectively. Caudal fin which is slender and terminates in a rounded leaf-like shape. Skin is densely covered by fine dermal denticles. It is purplish-brown to blackish above. Underside is white, with a narrow dark border along the lateral disk margins. Tail is entirely dark, and the caudal fin is black. This large species grows up to 2.7 m long, 1.5 m across, and 118 kg in weight (Nishida, 1990; Schwartz, 2007).

Food habits: Feeds on small pelagic fish, eels, crabs, shrimp, lobsters, cephalopods, and polychaete worms.

Reproduction: It is aplacental viviparous and developing embryos are nourished by maternally produced histotroph uterine milk. It has a small litter size and a long gestation period. Young are born when it is 50 cm. Males and females mature sexually at 1.3–1.7 m ft) and 1.9–2.0 m long, respectively.

Predators: Not reported.

Parasites: Cestoda: *Parachristianella monomegacantha*

IUCN conservation status: Least concern.

KEYWORDS

- marine stingrays
- common name
- distribution
- habitat
- species description
- food habits
- reproduction
- parasites

CHAPTER 4

BIOLOGY AND ECOLOGY OF FRESHWATER STINGRAYS

CONTENTS

ABSTRACT

The ecology, biology, and identifying features of freshwater stingrays along with their uses, threats, and conservation are given in this chapter.

4.1   ECOLOGY OF FRESHWATER STINGRAYS

4.1.1   DISTRIBUTION

The freshwater stingrays which are also commonly called "river sting-rays" belong to the family Potamotrygonidae and are completely adapted for living exclusively in freshwater environments. These stingrays are widely distributed in several river basins of the Neotropical region (South American region). Their chief adaptations are the ability to floating on the surface when the bottom water is poor in oxygen and the maternal care (de Carvalho et al., 2003). Compared to the marine stingrays, the life history and ecology of most freshwater and euryhaline stingrays are poorly known. The Amazon Basin has the most valid species of freshwater stingrays (13 species) and Brazil has the highest concentration of species (18 species). A summary of the range of distribution of the freshwater stingrays in Amazon Basin is given in the following table (de almeida et al., 2009; de Araújo et al., http://www.cites.org/common/com/ac/20/E20-inf-08.pdf).

4.1.2   HABITATS

The freshwater stingrays occupy diverse habitats in freshwater environ-ments, including beach sands, flooded forest, small creeks with mud or stone bottoms, and lakes. In all such habitats, they are the predators on the top of the food web. Like marine stingrays, they exhibit low fecundity, late maturation, and slow growth. However, the habitat reduction during low water periods makes the freshwater stingray species more vulnerable than their marine counterparts. Some freshwater stingrays are endemic and require very specific habitat conditions such as acid and poor oxygen water. They are normally restricted to waters where the salinity is normally less than 3 ppt. They exhibit unique physiological features such as the inability to retain urea, due to the absence of salt excretion by the rectal

Distribution of Species of Potamotrygonids in Amazon Basin.

Species	Basin or river drainage	Countries
Plesiotrygon iwamae	Amazon	Brazil, Ecuador, and Peru
Paratrygon aiereba	Amazon and Orinoco	Bolivia, Brazil, Ecuador, Peru, and Venezuela
Potamotrygon henlei	Amazon	Brazil
P. brachyura	Paraná, de La Plata, Uruguay, and Paraguay	Argentina, Brazil, Paraguay, and Uruguay
P. castells	Amazon, Paraná, Paraguay, and Guaporé	Argentina, Bolivia, Brazil, Paraguay, and Peru
P. constellata	Amazon	Brazil and Colombia
P. dumerilii	Amazon, Paraná, and Paraguay	Argentina, Brazil, Paraguay, and Uruguay
P. falkneri	Cuiabá, Paraná, and Paraguay	Argentina, Brazil, and Paraguay
P. hystrix	Paraná and Paraguay	Argentina, Brazil, Paraguay, and Uruguay
P. cf. hystrix	Negro	Brazil
P. humerosa	Amazon	Brazil
P. leopoldi	Amazon	Brazil
P. magdalenae	Magdalena and Atrato	Colombia
P. motoro	Amazon, Orinoco, and de La Plata	Argentina, Bolivia, Brazil, Colombia, French Guyana, Guyana, Paraguay, Peru, Surinam, and Uruguay
P. ocellata	Amazon	Brazil
P. orbignyi	Amazon, Orinoco, Paraná, and Paraguay	Argentina, Brazil, Bolivia, Colombia, French Guyana, Guyana, Paraguay, Peru, Surinam, Uruguay, and Venezuela
P. schroederi	Amazon and Orinoco	Brazil, Colombia, and Venezuela
P. schuemacheri	Paraná and Paraguay	Argentina, Brazil, and Paraguay
P. scobina	Amazon	Brazil
P. signata	Parnaíba	Brazil
P. yepezi	Maracaibo	Venezuela

gland and modifications in the ampullae of Lorenzini. Analysis of plasma concentration components in some species has shown a typical teleostean blood chemistry.

4.2   BIOLOGY OF FRESHWATER STINGRAYS

4.2.1   IDENTIFICATION OF FRESHWATER STINGRAYS

The freshwater stingrays are known for their polychromatism, that is, their body patterns and colors are extremely variable. The different species of freshwater stingrays are normally identified using the following characteristics:

1.   lateral markings on the tail,
2.   presence or absence of raised denticles along the tail,
3.   ratio of tail length to disk width, and
4.   size and position of the eyes.

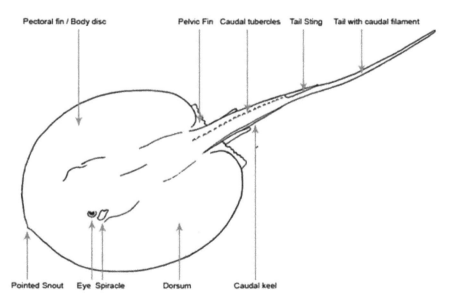

Body plan of a river stingray

4.2.2   CHARACTERISTICS OF FRESHWATER STINGRAYS (POTAMOTRYGONIDAE)

The river or freshwater stingrays are almost circular in shape and their size range from 25 cm (*Potamotrygon schuhmacheri*) to 1.5 m in diameter (*Potamotrygon brachyuran*). The dorsal surface of the disk and tail of these stingrays is usually covered with many denticles, thorns, and tubercles. The caudal sting (or serrated spine) located on the dorsal surface of the tail has small lateral serrations and an acute distal tip. These stings may be present up to four in one individual and possess longitudinal grooves to conduct venom produced in special glands at their bases. Interestingly, these stings are continuously worn, shed, and replaced. Additionally, many species have enlarged, nonvenomous spines on disk margins or over tail, sometimes in numerous rows. Oral teeth which are usually less than 50 rows in either jaw and set in quincunx are small with short cusps. They are however more prominent in adult males. Most potamotrygonid species have colorful dorsal arrangements, including spots of various dimensions, ocelli, reticulate patterns, and vermiform markings, which are generally species specific, and gray, brown, or black background coloration.

4.3   MORPHOLOGICAL CHARACTERISTICS OF DIFFERENT GENERA OF POTAMOTRYGONIDAE

The family of freshwater stingrays, namely, Potamotrygonidae, includes four genera, namely, *Potamotrygon*, *Paratrygon*, *Plesiotrygon*, and *Heliotryogn*. The chief characteristics of these genera are the following:

Potamotrygon: Species of this genus have moderately stout and short tails, usually shorter than disk length. The caudal sting is well developed and located farther posteriorly. Instead of dorsal and caudal fins, membranous skin folds (finfolds), with rudimentary internal radial elements occur on both upper and lower tail midlines. The eyes are moderately large.

Paratrygon: Species of this genus have slender, filiform, and whiplike tails. The tail is very long in juveniles but absolutely reduced in adults. The caudal sting is reduced and situated closer to the tail base. The disk is usually slightly longer than wide. The anterior margin of disk is concave, and the spiracle has a small protrusion. The finfolds are completely absent. The eyes are smaller and less protruded.

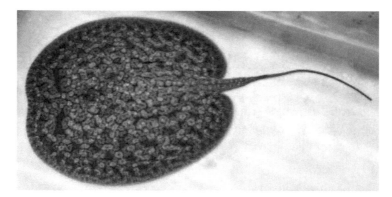

Plesiotrygon: Species of this genus have slender, filiform, and whiplike tails. Tail is not only much longer than disk length but also stout at base. The caudal sting is well developed and located farther posteriorly. Only ventral finfold is present. The anterior margin of disk is broadly pointed, and the spiracle is simple without protrusion. The eyes are smaller and less protruded. Large specimens may have more than 60 rows of teeth in either jaw.

Heliotrygon: This genus is known for its large size, pancake-like appearance, a distinct pattern of lateral line canals on the ventral surface, and a degenerate spine (http://www.elasmodiver.com/River_Stingrays_Potamotrygonidae.htm). It is a newly discovered genus with two species, namely, *Heliotrygon gomesi* and *Heliotrygon rosai*.

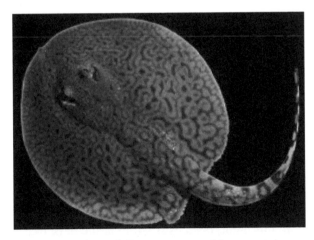

In addition to the species of Potamotrygonidae, certain species of the family Dasyatidae, namely, *Dasyatis laosensis*, *Dasyatis garouaensis*, and *Dasyatis ukpam*; *Himantura dalyensis*, *Himantura fluviatilis*, *Himantura kittipongi*, *Himantura krempfi*, *Himantura oxyrhyncha*, and *Himantura polylepis*; and a newly discovered genus, namely, *Makaraja* with its single species, namely, *Makararaja chindwinensis*, have also been reported to thrive in freshwater.

4.3.1   OSMOREGULATION

Unlike the euryhaline whiptail stingrays, riverine stingrays have completely lost the ability to migrate between freshwater and marine environments. Such a major modification to a fundamental function like osmoregulation (the regulation of internal solute concentrations) is quite remarkable in this group. Seawater is usually saltier than the blood of most fishes, but instead of actively pumping ions and other solutes out of their bodies like marine bony fishes, elasmobranchs simply match their internal osmotic concentrations to that of their external environment. They achieve this by maintaining the concentration of organic solutes (viz., urea and an enzyme called trimethylamine oxide—TMAO) within their bodies. Although urea is toxic to fish, the TMAO counteracts the protein-destabilizing effects of urea. Excess monovalent ions (viz., sodium and chloride) which they ingest are eliminated from the body via specialized rectal glands.

Since Amazonian rays live in freshwater, they have exactly the opposite problem of their marine counterparts. That is, instead of losing water to their external environment, they have to gain it, since their internal osmotic concentrations are higher than that of the water in which they live. As a result of this situation, for freshwater stingrays, there is no need for rectal glands, and these structures are now vestigial (greatly reduced in size and no longer capable of secreting salt). They have also lost the ability to retain urea.

4.3.2   EYESIGHT

The protruding "periscope" eyes of freshwater stingrays help them to see the water column above, while they are buried in the river bed. They also possess keen eyesight which allows them to navigate the murky environments in which they live. They are also adapted to low-light conditions in a special way. In dim light, ray's eyes shine in the dark like a cat. A layer located toward the back of the eye reflects light back into the retina, giving the ray night vision.

4.3.3   AMPULLAE OF LORENZINI

These jelly-filled sensory pores which are located on the skin around the nose and mouth on the underside of the disk help both marine and freshwater stingrays detect minute electric fields generated by other living organisms. This device is especially useful for these rays because it helps them to hunt down prey that might be buried in the riverbed or hiding in murky water.

4.3.4 SPIRACLES AND BREATHING

As the mouths and gills are located under their bodies in freshwater sting-rays, breathing is always a problem for these rays while they are on the sand. To overcome this hurdle, these rays have developed large spiracles which are positioned just behind the eyes. Through these openings, these rays can suck in oxygen-rich water to flush over the gills. Through this mechanism, river stingrays are able to remain motionless for hours (http:// www.elasmodiver.com/River_Stingrays_Potamotrygonidae.htm).

4.3.5 FOOD AND FEEDING

The adult forms of different species of freshwater stingrays feed mainly on fishes, worms, and small crustaceans and the juveniles feed on small crustaceans and aquatic insects. The food present in the sediment is sucked in. Like other marine stingrays, the tooth structure of freshwater sting-rays consists of small rounded molars which form a flat upper and lower surface designed to grip but not to cut.

4.3.6 LOCOMOTION

River stingrays are able to swim forward (and slowly backward) by undu-lating their pectoral fins which form their body disk. Their tails with their rudimentary caudal fins are used for steering and balancing and also to support their defensive tail stings.

4.3.7 DEFENSIVE MECHANISMS

River stingrays have one or more stingers (tail spines) on the top of their tails. These weapons are capable of puncturing the hide of a predator or impaling the leg or abdomen of a wading fisherman. Each stinger is sheathed in a

mildly venomous covering of skin which is pushed back as the point enters
the victim allowing the venom to come in contact with the cut tissue.

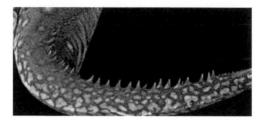

Dorsal view of tail showing pointed spines in *Potamotrygon motoro*.

4.3.8   REPRODUCTION

A stingray's sex can easily be determined based on the presence or absence
of claspers (two "penises") connected to the inside of the pelvic fins of
male rays. In the juvenile rays, they look like tiny nubs and are difficult
to identify. The eggs are fertilized internally after the male inserts one of
his claspers into the female's cloaca. The actual sex act lasts less than a
couple of seconds. The female may sometimes sustain slight injury during
mating when the male bites her. To counteract this, the female has evolved
thicker and more durable skin than the male (Góes de Araújo & Pinto de
Almeida, 2005).

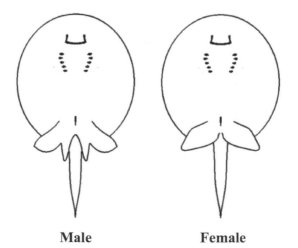

Male   Female

The freshwater stingrays are matrotrophically viviparous, giving birth to one to seven live young at a time after a gestation period of several months (depending on the species). The uterus is formed from the expansion of the oviduct. The embryos obtain nourishment from their yolk sacs early in their development. During the later stages of pregnancy, small, filamentous appendages called "trophonemata" develop within the uterus and penetrate the spiracle of the embryo supplying it with a nutrient-rich fluid called "histotrophe" until it is born. At birth, the young rays look like their parents and are able to search for their own food (Góes de Araújo & Pinto de Almeida, 2005). Gestation period observed in certain species of freshwater stingrays is given below.

Reproduction in some freshwater stingray species (Oldfield, 2005a,b).

Species	Gestation period (months)	Disk width[a] (mm)
Plesiotrygon iwamae	8	400/500
Paratrygon aierba	9	600/720
Potamotrygon motoro	6	390/440
P. orbignyi	6	390/440
P. schroederi	6	420/440
P. scobina	–	350/400
Potamotrygon sp.	3	160/170

[a]Male/Female.

4.3.9 FOOD VALUE

Among freshwater stingrays, *Potamotrygon magdalenae* is sometimes used for human consumption. When fished, species like *P. motoro* are often discarded as bycatch.

4.3.10 ORNAMENTAL VALUE

Certain species of Potamotrygonidae are much valued in ornamental fish trade. Species such as *Plesiotrygon iwamae, Paratrygon aiereba, Potamotrygon brachyuran, Potamotrygon castexi, P. constellata, P. dumerilii, P. falkneri, Potamotrygon henlei, Potamotrygon histrix, Potamotrygon* cf. *hystrix, P. humerosa, Potamotrygon leopoldi, Potamotrygon magdalenae,*

Potamotrygon motoro, Potamotrygon ocellata, P. orbignyi, P. schroederi, Potamotrygon schuemacheri, P. scobina, Potamotrygon signata, and *P. yepezi* are illegally transported from Brazil to international markets for ornamental purposes. As there are ample chances for illegal export of these fishes, it is time to regulate this ornamental fish trade.

4.3.11   THREATS

The following threats have been identified to the stingrays of the family Potamotrygonidae:

- subsistence/artisanal fisheries for food purposes,
- artisanal fisheries for ornamental purposes,
- commercial fisheries as bycatch,
- recreational fisheries,
- ecotourism (negative fisheries),
- habitat deterioration and destruction (including dragging, dams, and gold mines) may not only deplete freshwater stingray populations more severely, but they may also interfere with the life cycle of different stingray species (de Araújo et al., http://www.cites.org/common/com/ac/20/E20-inf-08.pdf).

4.3.12   RESEARCH AND CONSERVATION PRIORITIES

The biological data relating to freshwater stingrays such as distribution, population size and dynamics, habitat requirements, reproductive biology, life history, current levels of exploitation, and rates of anthropogenic habitat erosion or loss are virtually unknown. The biologists often encounter difficulties in identifying potamotrygonid stingrays, and the same is also with the researchers working on the medical aspects of stingrays. Further, fish traders and breeders are usually less concerned with the taxonomy of the freshwater stingrays. Therefore, consolidated efforts need to be taken to offer intensive training programs among biologists and medical professionals (de Araújo et al., http://www.dfo-mpo.gc.ca/Library/315632.pdf; Ramírez & Davenport, 2013). The conservation status of most freshwater and euryhaline stingrays has not been determined. So far, only five species of freshwater stingrays have been cited in the IUCN Red List as threatened species. This is largely due to a combination of taxonomic problems and absence of the data absolutely vital to developing informed management or conservation strategies for these fish. Many freshwater stingrays, due to their inherently low reproductive potential, may be driven to extinction as bycatch of fisheries. Further, some freshwater stingray populations may be decimated within a few decades, underscoring the importance of not wasting time duplicating research that may already be completed for a given species under synonyms.

KEYWORDS

- **freshwater stingrays**
- **maternal care**
- **food web**
- **polychromatism**
- **denticles**
- **threats**
- **conservation**

PROFILE OF FRESHWATER STINGRAYS

CONTENTS

ABSTRACT

The biology, ecology, and parasites of freshwater stingray species of the different families are given in this chapter.

5.1 RIVERINE STINGRAYS (POTAMOTRYGONIDAE)

5.1.1 *POTAMOTRYGON BRACHYURA* (Günther, 1880)

Phylum: Chordata   Subphylum: Vertebrata
Class: Chondrichthyes   Subclass: Elasmobranchii
Order: Myliobtiformes   Family: Potamotrygonidae

Common name: Giant freshwater stingray; short-tailed river stingray.

Geographical distribution: Temperate; South America: Paraná-Paraguay (including Cuiabá river in Brazil), Uruguay basins, and Argentina.

Habitat: Freshwater; demersal; very shallow, quiet waters of lagoons, brooks, and streams.

Distinctive features: These rays are circular in shape and humped in the back. Its disk is much thicker and its tail is much, much shorter, and more muscular. It has a brown, chain pattern over a slightly lighter brown background. Tail is covered with short spines at the base and a venomous sting at the end. Maximum size and weight of this species are 95 cm DW and 300 kg, respectively.

Food habits: Juveniles feed on small molluscs (lamellibranchs and gastropods), crustaceans, larvae of aquatic insects, and fish.

Reproduction: This species is matrotrophically viviparous. Embryos obtain nourishment from their yolk sacs initially. During the later stages of pregnancy, small, filamentous appendages called trophonemata develop within the uterus and penetrate the spiracle of the embryo supplying it with a nutrient-rich fluid called histotrophe. Gravid females are usually over 40 cm DW. Only the left ovary is functional. Maximum litter size is 19 pups. Pups normally feed on plankton after birth (Charvet-Almeida et al., 2009).

Predators: Major predators of this species are piranhas and humans.

Parasites

Monogenea: *Potamotrygonocotyle chisholmae*

P. dromedaries

Potamotrygonocotyle uruguayensis

Cestoda: *Rhinebothrium paratrygoni*

IUCN conservation status: Data deficient.

5.1.2   *POTAMOTRYGON CONSTELLATA* (Vaillant, 1880)

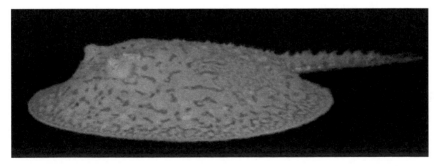

Common name: Thorny river stingray.

Geographical distribution: Tropical; South America: Amazon, and Solimões River basin.

Habitat: Freshwater; benthopelagic; potamodromous; prefers silty or sandy bottom of rivers and streams.

Distinctive features: Color of disk of this species is brown or dark gray with small white or yellow spots forming rosettes near the margins of the disk. A reticular pattern of dark pigment may be seen occasionally. Pelvic fins are with the same pattern dorsally as the disk. Disk is brown dorsally, with dark lateral stripes and ventrally mottled with irregular white spots. Upper jaw teeth are small and sharp. Anterior caudal half has thorns. This species grows to a maximum size and weight of 62 cm DW and 10 kg.

Food habits: Feeds primarily on other fish.

Reproduction: This species is matrotrophically viviparous. Embryos obtain nourishment from their yolk sacs early in their development. During the later stages of pregnancy, small, filamentous appendages called trophonemata develop within the uterus and penetrate the spiracle of the embryo supplying it with a nutrient-rich fluid called histotrophe. Males of this species reach first maturity age when its DW is half size the maximum DW.

Predators: Major predators of this species are piranhas and humans.

Parasites

Monogenea: *Paraheteronchocotyle amazonense*

Potamotrygonocotyle tsalickisi

Cestoda: *Acanthobothrium amazonensis*

Potamotrygonocestus amazonensis

Potamotrygonocestus travassosi

Rhinebothroides circularisi

IUCN conservation status: Data deficient.

5.1.3 *POTAMOTRYGON AMANDAE* (Loboda & de Carvalho, 2013)

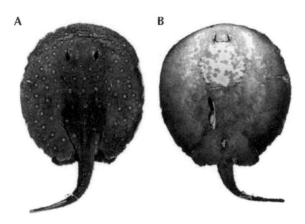

A. Dorsal view   B. Ventral view

Common name: River pooch.

Geographical distribution: Tropical; South America: Paraná-Paraguay basin.

Habitat: Freshwater; benthopelagic.

Distinctive features: This species has a predominantly grayish or dark brown dorsal background color usually with bicolored ocelli on dorsal disk. A whitish, light gray, or light yellow central area surrounded by a black peripheral ring is also present. Ventral disk is grayish. This species grows to a maximum size of 31.2 cm (DW) (Loboda & de Carvalho, 2013).

Food habits: It feeds on shrimp, crab, snails, insect larvae, and other small animals.

Reproduction: Not much is known about the biology of this species. However, like other freshwater stingrays, it is matrotrophically viviparous. The embryos obtain nourishment from their yolk sacs early in their development. During the later stages of pregnancy, small, filamentous appendages called trophonemata develop within the uterus and penetrate the spiracle of the embryo supplying it with a nutrient-rich fluid called histotrophe.

Predators: Major predators of this species are piranhas and humans.

Parasites: Not reported.

IUCN conservation status: Not evaluated.

5.1.4 *POTAMOTRYGON BOESEMANI* (Rosa et al., 2008)

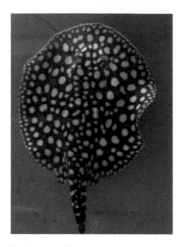

Common name: Not designated.

Geographical distribution: Tropical; South America: Corantijn river drainage in Suriname.

Habitat: Freshwater; benthopelagic.

Distinctive features: Dorsal disk of this species has dark brown background coloration. Further, deep-orange to red ocellated spots, encircled by relatively broad black rings, are also present. Buccal cavity is dark-pigmented with orange spots in adults. It grows to a maximum length of 41.3 cm DW (Rosa et al., 2008).

Food habits: It feeds on shrimp, fish fillet, worms, and other common foods.

Reproduction: Not much is known about the biology of this species. However, like other freshwater stingrays, it is matrotrophically viviparous. The embryos obtain nourishment from their yolk sacs early in their development. During the later stages of pregnancy, small, filamentous appendages called trophonemata develop within the uterus and penetrate the spiracle of the embryo supplying it with a nutrient-rich fluid called histotrophe.

Predators: Major predators of this species are piranhas and humans.

Parasites: Not reported.

IUCN conservation status: Not reported.

5.1.5 *POTAMOTRYGON CASTEXI* (Castello & Yagolkowski, 1969).

Common name: Vermiculate river stingray.

Geographical distribution: South America inland waters: Paraná and Paraguay River basins in Argentina, Paraguay, and Brazil.

Habitat: Silty or sandy bottom of rivers and streams.

Distinctive features: Tail of this species is powerfully built and is longer than the body. Upper surface of tail is irregularly light spotted to a dark background or covered in dark lines on a lighter background. Lower half of the tail appears striped or barred when viewed laterally. Body disk has a fine light edge. Base color of the disk is light to dark brown. This species grows to a maximum size of 60 cm DW (https://rybicky.net/atlasryb/trnucha_castexova).

Food habits: Feeds primarily on fish.

Reproduction: Not much is known about the biology of this species. However, like other freshwater stingrays, it is matrotrophically viviparous. Embryos obtain nourishment from their yolk sacs early in their development. During the later stages of pregnancy, small, filamentous appendages called trophonemata develop within the uterus and penetrate the spiracle of the embryo supplying it with a nutrient-rich fluid called histotrophe.

Predators: Major predators of this species are piranhas and humans.

Parasites

Cestoda: *Anindobothrium guariticus*

Potamotrygonocestus sp.

Rhinebothrium sp.

Rhinebothroides sp.

IUCN conservation status: Data deficient.

5.1.6 *POTAMOTRYGON FALKNERI* (Castex & Maciel, 1963)

Common name: Large spot stingray.

Geographical distribution: Rio Paraná and Rio Paraguay basins in Brazil, Paraguay, and Argentina.

Habitat: Sand banks, shallows of major rivers, and slow-moving tributaries with substrates of mud or sand; flooded forests; terrestrial lakes and ponds formed by the receding flood waters.

Distinctive features: Disk of this species is oval and only slightly longer than wide. Anterior margin of disk is convex with a small fleshy protuberance on snout. Posterior margins of disk are also convex and their inner margins are fused posteriorly to the dorsal surface of pelvic fins and tail base. Disk is dorsoventrally compressed and slender. Anterior portion of disk is with small, prominent, and oval-shaped eyes. Spiracles are oval and small. Mouth is small and mouth opening is relatively straight across and with five buccal papillae. Labial ridges are present. Disk has a light brown background and has uniformly yellow dots or even oval or kidney-shaped spots. It has no tail spines and a relatively short tail. Large females of this species achieve a disk size of 47 cm diameter (da Silva & de Carvalho, 2011).

Food habits: Feeds chiefly on other fish and aquatic invertebrates, including worms and crustaceans.

Reproduction: This species shows matrotrophically viviparous. Embryos obtain nourishment from their yolk sacs early in their development. During the later stages of pregnancy, small, filamentous appendages called trophonemata develop within the uterus and penetrate the spiracle of the embryo supplying it with a nutrient-rich fluid called histotrophe. After a gestation period of about 20 weeks, two pups are born (Compagno, 1999).

Predators: Major predators of this species are piranhas and humans.

Parasites

Monogenea: *Potamotrygonocotyle eurypotamoxenus*

Potamotrygonocotyle tsalickisi

Cestoda: *Acanthobothrium regoi*

Nandocestus guariticus

Paroncomegas araya

Eutetrarhynchus araya

Potamotrygonocestus amazonensis

Potamotrygonocestus travassosi

Rhinebothrium paratrygoni

Ribautia paranaensis

Trematoda: *Clinostomum complanatum*

Nematoda: *Brevimulticaecum* sp.

Cucullanus (*Cucullanus*) sp.

Echinocephalus sp.

Spinitectus sp.

Acanthocephala: *Quadrigyrus machadoi*

IUCN conservation status: Data deficient.

5.1.7 *POTAMOTRYGON HENLEI* (Castelnau, 1855)

Common name: Bigtooth river stingray, fire ray.

Geographical distribution: Tropical; South America: Tocantins River basin (Tocantins and Araguaia rivers).

Habitat: Freshwater; benthopelagic; prefers mud bottoms.

Distinctive features: This species has a light brown to deep-black background. Dorsal disk of this species has generally larger spots, particularly around the perimeter of the disk. These big spots are surrounded by a series of smaller white spots giving the appearance of a crown when viewed from above. Color of eye is brown. This species grows to a maximum DW and total length (TL) of 71 and 104 cm, respectively.

Food habits: It feeds mainly on gastropod molluscs.

Reproduction: It is ovoviviparous. Females mature at a TL of 80–85 cm (around 50 cm DW) and number of young ranges from 1 to 9. Size at birth is approximately 25–30 cm TL. Gestation time and reproductive periodicity are unknown. It breeds year-round. Pups may have dots on the dorsal side of the disk like their parents.

Predators: This species has few natural predators (http://animals.pawnation.com/bigtooth-stingrays-2890.html).

Parasites

Monogenea: *Potamotrygonocotyle chisholmae*

Potamotrygonocotyle dromedarius

Cestoda: *Rhinebothrium copianullum*

Pharmacological properties of mucous and venom: The mucus and sting venom of this species have been reported to be toxic to mice with nociceptive, edematogenic, and proteolysis activities. Further, the venom of this species possesses a diverse mixture of peptides, enzymes and pharmacologically active components (Conceição et al., 2012; Monteiro-dos-Santos et al., 2011).

IUCN conservation status: Least concern.

5.1.8 *POTAMOTRYGON HYSTRIX* (Müller & Henle, 1841)

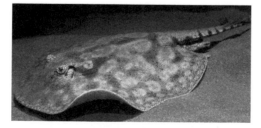

Common name: Porcupine river stingray.

Geographical distribution: Subtropical; South America: Paraná-Paraguay river basin.

Habitat: Freshwater; benthopelagic; marshy zones where it is found partially hidden in the sandy bottom.

Distinctive features: This species has a tail which is equipped with one or more deciduous spines. Spine which is barbed has a length of 4–6 cm and is inserted dorsally in the middle portion of the tail. Spine is coated with an extremely toxic mucus produced by the cells of the skin and inflicts very painful wounds. It has a maximum size and weight of 40 cm DW and 15 kg, respectively (Compagno, 1999).

Food habits: Juveniles feed on plankton and adults on small molluscs, crustaceans (crabs), insect larvae, small fish, and other food items.

Reproduction: This species shows matrotrophically viviparous. Embryos obtain nourishment from their yolk sacs early in their development. During the later stages of pregnancy, small, filamentous appendages called trophonemata develop within the uterus and penetrate the spiracle of the embryo supplying it with a nutrient-rich fluid called histotrophe. Females have been reported to give birth to 6–9 pups.

Predators: Carnivorous predators like larger fish and caiman.

Parasites

Monogenea: *Potamotrygonocotyle eurypotamoxenus*

Potamotrygonocotyle tsalickisi

Cestoda: *Acanthobothrium regoi*

Paroncomegas araya

Eutetrarhynchus araya

Potamotrygonocestus travassosi

Rhinebothroides freitasi

Rhinebothroides glandularis

Rhinebothroides scorazai

Rhinebothroides venezuelensi

Rhinebothrium paratrygoni

IUCN conservation status: Data deficient.

5.1.9 POTAMOTRYGON ITAITUBA, Christian Höhne

Common name: Polka-dot river stingray, spotted river ray, galaxy river ray.

Geographical distribution: Rio Tapajos, Xingu River Basin, Itaituba, Brazil.

Habitat: Rivers; benthopelagic.

Distinctive features: This species has a differentiated pattern and extra spotting around the disk margin. It has a dark brown to black color with a multitude of small white to pale yellow ocelli dorsally. Underside of the disk is mottled. On the tail, it has 5–7 rows of thorns. This species grows to a maximum size of 50 cm DW (http://rochen.chapso.de/potamotrygon-itaituba-s235039.html).

Food habits: It thrives on live foods.

Reproduction: It is presumed to be ovoviviparous. Not much is known about the biology or life history of this species.

Predators: Carnivorous predators like larger fish and caiman.

Parasites: Not reported.

IUCN conservation status: Not available.

5.1.10   *POTAMOTRYGON LEOPOLDI* (Castex & Castello, 1970)

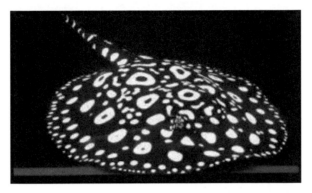

Common name: White-blotched river stingray, Xingu river ray, black diamond.

Geographical distribution: Tropical; South America: Xingu river basin (Xingu and Fresco rivers).

Habitat: Freshwater; benthopelagic; buried in sand during day.

Distinctive features: It is one of the most stunning freshwater stingray species. Dorsal surface of the disk of this species is black to very dark brownish-gray. White round, lunate or circular spots are present. Belly is light gray. Several rows of spines are seen on dorsal surface of the tail. It can cause severe injuries with the venomous dentine spine located at its tails. This species has a maximum DW of 40 cm; TL of 80 cm and weight of 25 kg (Kirchhoff et al., 2014).

Food habits: Hunts for benthic invertebrates during night; wild rays normally feed on other fish and invertebrates such as crustaceans and worms. Small fish and shrimp are also good meals of this species.

Reproduction: This species shows matrotrophically viviparous. Embryos obtain nourishment from their yolk sacs early in their development. During the later stages of pregnancy, small, filamentous appendages called trophonemata develop within the uterus and penetrate the spiracle of the embryo supplying it with a nutrient-rich fluid called histotrophe. It generally gives birth after 12–15 weeks, depending on water temperature and other factors. It has a relatively high level of fertility compared to other potamotrygonidés, ranging from 4 to 12 young per litter (average of 7–8). The pups usually have a small yolk sac attached at birth, and they will feed from this up to a week. After the sac has been absorbed they should be offered high quality live and frozen foods several times a day.

Predators: Other stingrays or large catfishes.

Parasites

Monogenea: *Potamotrygonocotyle chisholmae*

Potamotrygonocotyle dromedarius

Cestoda: *Potamotrygonocestus fitzgeraldae*

Rhinebothrium copianullum

Rhinebothroides freitasi

IUCN conservation status: Data deficient.

5.1.11 *POTAMOTRYGON LIMAI* (Fontenelle et al., 2014)

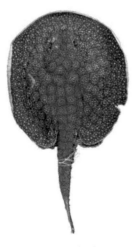

A. Dorsal view   B. Ventral view

Common name: Not designated.

Geographical distribution: Tropical; South America: Jamari river, upper Madeira river basin in Brazil.

Habitat: Freshwater; benthopelagic.

Distinctive features: Dorsal disk of this species has a dark brownish background, covered with whitish, closely packed small spots which are arranged in small concentric patterns. Ocelli are absent. Lower back portion of disk is with a characteristic polygonal pattern. Rostral dermal denticles are relatively simple and are composed of a single central crown and star-shaped base. Central and posterior disk denticles possess star-shaped crown ridges. Two to three irregular rows of hook-shaped spines are present on dorsal tail midline. This species grows to a maximum size of 64.8 cm DW (Fontenelle et al., 2014)

Food habits: Normally feeds on other fish and invertebrates such as crustaceans and worms.

Reproduction: This species shows matrotrophically viviparous. Embryos obtain nourishment from their yolk sacs early in their development. During the later stages of pregnancy, small, filamentous appendages called trophonemata develop within the uterus and penetrate the spiracle of the embryo supplying it with a nutrient-rich fluid called histotrophe.

Predators: Not reported.

Parasites: Not reported.

IUCN conservation status: Not reported.

5.1.12 *POTAMOTRYGON MAGDALENAE* (Duméril, 1865)

Common name: Magdalena river stingray.

Geographical distribution: Tropical; South America: Magdalena and Atrato river basins.

Habitat: Freshwater; benthopelagic; prefers shallow muddy bottoms with turbid waters.

Distinctive features: This species is flattened dorsoventrally. In the ventral surface, it has five pairs of gills. Mouth is small, transverse, and tail ends in filament, with lateral longitudinal folds at the base and the distal position is with a serrated spine. Body color of this species is light to dark brown and has several yellow ocelli (eyespots). Disk rim is usually smaller. This species is relatively small and has a maximum size of only 35.0 cm DW.

Food habits: Feeds mainly on insect larvae.

Reproduction: This species shows matrotrophically viviparous. The embryos obtain nourishment from their yolk sacs early in their development. During the later stages of pregnancy, small, filamentous appendages called trophonemata develop within the uterus and penetrate the spiracle of the embryo supplying it with a nutrient-rich fluid called histotrophe. This species reaches maturity with less than 25 cm of DW and has low fecundity (Charvet-Almeida & de Almeida, 2009).

Predators: It has relatively few predators, except for some larger fish and caiman.

Parasites

Cestoda: *Acanthobothrium quinonesi*

Potamotrygonocestus magdalenensis

P. moralarai

Rhinebothroides moralarai

Trematoda: *Paravitellotrema overstreeti*

IUCN conservation status: Near threatened.

5.1.13 *POTAMOTRYGON MARINAE* (Deynat, 2006)

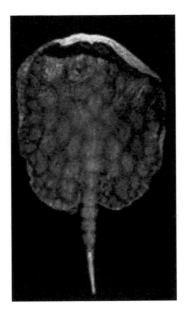

Common name: River stingray.

Geographical distribution: Tropical; South America: French Guyana (Oyapock and Maroni rivers).

Habitat: Freshwater; benthopelagic.

Distinctive features: This species is distinguished by its feebly prepelvic process, postorbital process (an enlarged blade), and unsegmented angular cartilage surface of the dorsal coloration which is composed of wide circular patches. Further, it has dark coloration in the ventral surface which is tessellated with pale patches. Small-sized spiny tubercles are found situated in the middorsal area, before the caudal sting. The largest specimen measures 41.2 cm DW and juveniles measure 23.8–29.8 cm DW (Comptes Rendus Biologies, 2006; Deynat, 2006; Valenti, 2009b).

Nothing is known about its Biology.

IUCN conservation status: Data deficient.

5.1.14 *POTAMOTRYGON MOTORO* (Müller & Henle, 1841)
(=Potamotrygon laticeps, Potamotrygon circularis)

Common name: Ocellate river stingray, South-American freshwater stingray, motoro stingray

Geographical distribution: It has a widespread distribution, extending across Argentina, Brazil, Paraguay, Uruguay, and Venezuela.

Habitat: Restricted entirely to freshwater; favors calm waters, especially the sandy edges of lagoons, brooks, and streams, often found lying still, buried in the sandy sediment at the bottom of a stream, particularly during the warmest part of the day.

Distinctive features: Disk of this species is subcircular and slightly longer than broad. Head is large, with relatively wide interorbital space. Mouth is relatively large, presenting five buccal papillae. Eyes are bulging dorsally and relatively large. Spiracles are muscular, relatively large, and trapezoidal. Teeth are arranged in quincunx and are relatively large. In adults, teeth are sexually dimorphic. Males are with cusps on central row of both jaws and females with flattened teeth in all rows. Pelvic fins are generally covered dorsally by disk or protruding only slightly. Lateral margin is slightly rounded and posterior margin is undulated. Anterior margins of pelvic fins are oblique to midline. Claspers are relatively conical and are slightly tapered posteriorly. Tail is thick and moderately short. Dorsal disk background is gray, dark gray, olive, olivaceous brown, or dark brown, with ocelli distributed over entire disk to base of tail. Ventral disk coloration is divided into two regions, one lighter colored at disk center, usually whitish, light yellow, or beige, and another on outer disk periphery with a darker gray or light brown color. Maximum DW is 100 cm and weight is 15 kg.

Food habits: Initially after birth, this species feeds on plankton, but as it grows, the diet expands to include small molluscs, crustaceans, and the larvae of aquatic insects, while larger adults also eat catfishes of the family Loricariidae.

Reproduction: Females of this species produce eggs, but these develop inside the female. Young hatch inside the female and are then born live. Sexual maturity for male and female has been recorded at a DW of 390 and 440 mm, respectively. Gonadal maturation and gestation period are at 3–4 and 6 m, respectively, and each female gives birth to 6–11 young.

Predators: Humans and amphibious carnivores and alligatorid crocodilians are the major predators of this species.

Parasites

Monogenea: *Potamotrygonocotyle auriculocotyle*

Potamotrygonocotyle chisholmae

P. dromedaries

Potamotrygonocotyle eurypotamoxenus

P. rionegrense

Potamotrygonocotyle tsalickisi

Cestoda: *Acanthobothrium peruviense*

A. ramiroi

A. terezae

Paroncomegas araya

Eutetrarhynchus araya

Potamotrygonocestus amazonensis

Potamotrygonocestus fitzgeraldae

Potamotrygonocestus travassosi

Rhinebothroides campbelli

R. mclennanae

R. scorazai

Rhinebothroides sp.

Rhinebothrium corbatai

R. mistyae

Rhinebothrium paratrygoni

Nematoda: *Brevimulticaecum regoi*

Acanthocephala: *Quadrigyrus machadoi*

IUCN conservation status: Not considered an endangered species.

5.1.15 *POTAMOTRYGON OCELLATA* (Engelhardt, 1912)

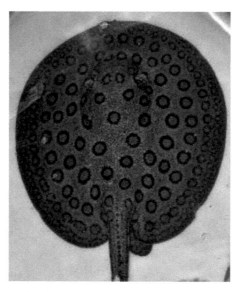

Common name: Red-blotched river stingray.

Geographical distribution: Tropical; South America: Pedreira river and south of Mexiana Island in Brazil.

Habitat: Freshwater; benthopelagic.

Distinctive features: This species has dark red to brown rust ocelli and a brown to olive-brown color. Some ocelli may have a dark edge. Eyespots are absent on the tail. One row of dorsal tail spines is present. This species reaches a maximum size of 20 cm DW. Values of maximum TL and weight are 110 cm and 44 kg, respectively.

Food habits: It may feed on a variety of live and frozen foods. Very little is currently known of the life history characteristics of this species.

Reproduction: Males have the modified pelvic fins which they use to fertilize the female. The latter gives birth to a batch of puppies (ray fry).

IUCN conservation status: Data deficient.

5.1.16 *POTAMOTRYGON ORBIGNYI* (Castelnau, 1855)
(=Potamotrygon humerosa, Potamotrygon dumerilii, Potamotrygon reticulates)

Common name: Smooth back river stingray.

Geographical distribution: Tropical; South America: Amazon and Orinoco river basins; river systems in Suriname, Guyana, and French Guiana.

Habitat: Freshwater; benthopelagic; stream swamps, floodplain lakes, and artificial lagoons.

Distinctive features: Disk of this species is rough with sharp denticles. Tail is with median dorsal and lateral spines which are enlarged and irregularly distributed from the base. Coloration of the dorsal disk of this species is variable. Upper surface of the disk and pelvic fins is brown dark or light brown, usually with a reticular pattern of dark pigment and circular or hexagonal spaces. Tail is brown dorsally, with dark sidebars and ventrally mottled with irregular white spots. This species has a maximum size of 60 cm DW and weight of 10 kg.

Food habits: Feeds mainly on insects and benthic crustaceans.

Reproduction: Females of this species produce eggs, but these develop inside the female. Young hatch inside the female and are then born live. For this species, sexual maturity for male and female was recorded at a DW of 390 and 440 mm, respectively. Gonadal maturation and gestation period were in 3–4 and 6 m, respectively, and each female gives birth to 3–7 young (http://www.researchgate.net/profile/Joao_Paulo_Da_Silva4/ publication/270760341_Familia_Potamotrygonidae_-_Potamotrygon_ humerosa/links/54b41b880cf28ebe92e4535b.pdf; Lasso et al., 1997).

Toxin: A new class of fish toxins, namely, porflan (natural peptide), has been isolated from this species and the same has been reported to be directly involved in the inflammatory processes.

Predators: Humans and caimans.

Parasites

Monogenea: *Potamotrygonocotyle chisholmae*

P. rionegrense

Potamotrygonocotyle tsalickisi

Cestoda: *Anindobothrium lisae*

Paroncomegas araya

Eutetrarhynchus araya

Paroncomegas baeri

Potamotrygonocestus amazonensis

Potamotrygonocestus fitzgeraldae

Potamotrygonocestus maurae

P. orinocoensis

Potamotrygonocestus travassosi

Rhinebothroides glandularis

R. scorazai

Rhinebothrium brooksi

Rhinebothrium copianullum

R. fulbrighti

Rhinebothrium paratrygoni

Copepoda: *Ergasilus trygonophilus*

IUCN conservation status: Least concern.

5.1.17 POTAMOTRYGON PANTANENSIS (Loboda & de Carvalho, 2013)

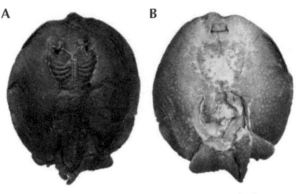

A. Dorsal view   B. Ventral view

Common name: River pooch.

Geographical distribution: Tropical; South America: northern Pantanal region, Brazil.

Habitat: Freshwater; benthopelagic.

Distinctive features: This species possesses bicolored ocelli with a size greater or equal to that of its eye. Background color of dorsal disk is uniform brown with yellow, vermiculated markings. Further, a single, clearly demarcated gray color is present on anterocentral ventral disk, transversed by a gray stripe over first pair of branchial slits. Minute, star-shaped dermal denticles are present only on central disk area, and two or three rows of minute, slender and curved enlarged spines on dorsal tail. This species grows to a maximum size of 26.8 cm DW (Loboda & de Carvalho, 2013).

Nothing is known about the biology of this species.

5.1.18   *POTAMOTRYGON   SCHROEDERI* (Fernández-Yépez, 1958)

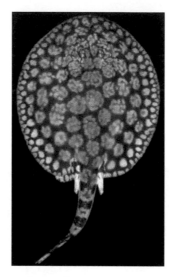

Common name: Rosette river stingray.

Geographical distribution: Tropical; South America: Apure (Orinoco basin) and Negro (Amazon basin) rivers.

Habitat: Freshwater; benthopelagic; juveniles inhabit shallow water around sandy beaches, and small creeks; adults occupy main river channels and sandy beaches.

Distinctive features: Lateral markings on the tail and the presence or absence of raised denticles along the tail are the characteristic features of this species. Individuals may exhibit clusters of spots which form flowerlike patterns on a plain or subtly mottled background. Flower patterns become smaller and more uniformly circular toward the edge of the disk. Some specimens may display a dark reticulated pattern enclosing irregularly shaped light spots. There are endless intermediate color morphs. Tail is always patterned with distinct and regularly spaced dark bars. Maximum DW is 60 cm (Góes de Araújo, 2009).

Food habits: Food includes small fishes, shrimps, worms, and insect larvae.

Reproduction: Reproductive mode is matrotrophic viviparity with trophonemata. Reproductive cycle is annual. Females of this species produce eggs, but these develop inside the female. Young hatch inside the female and are then born live. For this species, sexual maturity for male and female was recorded at a DW of 420 and 440 mm, respectively. Gonadal maturation and gestation period were in 3–4 and 6 m, respectively, and each female gives birth to 3–7 young. Size at birth is 14 cm DW.

Predators: Humans and caimans.

Parasites: Monogenea: *Potamotrygonocotyle rarum.*

IUCN conservation status: Data deficient.

5.1.19 *POTAMOTRYGON SCHUHMACHERI* (Castex, 1964)

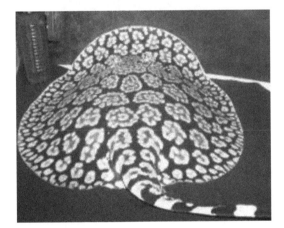

Common name: Parana river stingray

Geographical distribution: Temperate; South America: Paraná-Paraguay river basin.

Habitat: Freshwater; benthopelagic

Distinctive features: Dorsal surface of the disk is covered with denticles. This venomous species can provoke severe envenomations in fishermen, bathers, and riverside inhabitants of freshwater environments. The injuries due to this species may cause intense local pain and cutaneous necrosis, complicated by secondary infections and retention of fragments of the stingers in the wound. It has a maximum size of 25 cm DW.

Nothing is known about the biology of this species.

5.1.20 *POTAMOTRYGON SCOBINA* (Garman, 1913)

Common name: Raspy river stingray.

Geographical distribution: Tropical; South America: Middle and lower Amazon River, lower Tocantins river, Pará river, Trombetas river in Brazil.

Habitat: Freshwater; benthopelagic; potamodromous.

Distinctive features: Basic color of this species is light brown, dark brown, or even greenish. Ocelli are large and edge of the disk has many small white to yellow dots. Values of maximum DW, TL, and weight reported are 69 cm, 133 cm, and 15 kg, respectively.

Food habits: Food items of this species include mainly isopods (Sphaeromatidae) and shrimps (Paleomonidae).

Reproduction: Reproductive mode is matrotrophic viviparity with trophonemata. Reproductive cycle is annual. Females of this species produce eggs, but these develop inside the female. Young hatch inside the female and are then born live. Sexual maturity for male and female is at a DW of 350 and 400 mm, respectively. Gestation period is 6 m and each female gives birth to 1–16 young. Size at birth is 14 cm DW.

Predators: Humans and caimans.

Parasites

Monogenea: *Potamotrygonocotyle auriculocotyle*

Potamotrygonocotyle chisholmae

Potamotrygonocotyle septemcotyle

Potamotrygonocotyle tocantinsense

Cestoda*: Potamotrygonocestus amazonensis*

Copepoda: *Ergasilus trygonophilus*

IUCN conservation status: Data deficient.

5.1.21 *POTAMOTRYGON SIGNATA* (Garman, 1913)

Common name: Parnaiba river stingray.

Geographical distribution: Tropical; South America: Parnaíba river basin.

Habitat: Freshwater; benthopelagic.

Distinctive features: This species has a brown to reddish skin and white to yellowish ocelli with dark borders. Ocelli are small and look like speckled. Adult animals may possess 2–3 rows of tail spines. Maximum size of this species is 30 cm DW. The life history aspects of this species are completely unknown.

Food habits: It is an insectivorous species and its food items include Diptera larvae (60.64%) and Ephemeroptera nymphs (34.68%) (Moro et al., 2012).

Reproduction: Reproductive mode is matrotrophic viviparity with trophonemata. Reproductive cycle is annual. Females of this species produce eggs, but these develop inside the female. Young hatch inside the female and are then born live (Rosa et al., 2009).

Predators: Humans and caimans.

Parasites: Not reported.

IUCN conservation status: Data deficient.

5.1.22 *POTAMOTRYGON TATIANAE* (da Silva & de Carvalho, 2011)

Common name: River stingray.

Geographical distribution: Tropical; South America: known only from Río Madre de Díos, upper Río Madeira basin, Peru.

Habitat: Freshwater; demersal.

Distinctive features: Dorsal disk of this species has a dark background with a beige or light brown, closely packed, and highly convoluted vermicular pattern. One row of irregular spines is present on dorsal tail midline. Star-shaped, asymmetrical, and minute dermal denticles are present. Eyes are small. Mouth is small and it has two lateral and three central papillae.

One of the central papillae is closer to the lower jaw tooth plate. Teeth are relatively small, with 36–46 rows in the upper jaw and 33–45 in the lower jaw. It grows to a maximum DW of 36.2 cm and TL of 75.5 cm (da Silva & de Carvalho, 2011).

Nothing is known about its biology (da Silva & de Carvalho, 2011).

5.1.23 *POTAMOTRYGON TIGRINA* (de Carvalho et al., 2011)

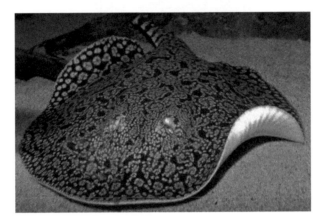

Common name: Tiger ray.

Geographical distribution: Tropical; South America: Río Nanay, upper Amazon basin in Peru.

Habitat: Freshwater; benthopelagic.

Distinctive features: This species is beautifully colored and is identified by its conspicuous dorsal disk coloration which is composed of bright yellow to orange vermiculations strongly interwoven with a dark-brown to deep-black background. A single angular cartilage is present. Dorsal tail spines are low and are not closely grouped. Coloration of tail is composed of relatively wide and alternating bands of creamy white and dark brown to black. This species grows to a maximum DW of 50 (male) and 80 cm (female). This species is frequently commercialized in the international aquarium trade (de Carvalho et al., 2011; de Carvalho et al., http://www.producao.usp.br/handle/BDPI/27625).

Virtually, nothing is known of its biology or conservation status.

5.1.24 *POTAMOTRYGON YEPEZI* (Castex & Castello, 1970)

Common name: Maracaibo river stingray.

Geographical distribution: Tropical; South America: Rivers draining into Maracaibo Lake.

Habitat: Freshwater; benthopelagic; prefers shallow muddy bottoms with turbid waters.

Distinctive features: This species is identified by the indistinct black splotches and streaks present on a light brown base color. Occasionally, lighter spots are present around the edges of the disk. Markings on the sides of the tail are also unevenly distributed and blotchy. This species has a maximum size of 40.0 cm DW (http://www.raylady.com/Potamotrygon/species/Yepezi.html).

Food habits: Feeds on insect larvae.

Reproduction: Reproductive mode is matrotrophic viviparity with trophonemata. Reproductive cycle is annual. Females of this species produce eggs, but these develop inside the female. Young hatch inside the female and are then born live.

Predators: Not reported.

Parasites

Cestoda: *Acanthobothrium quinonesi*

Potamotrygonoceslus amazonesis

Rhinebothroides venezuelensis

IUCN conservation status: Data deficient.

5.1.25 *HELIOTRYGON GOMESI,* de Carvalho & Lovejoy, 2011

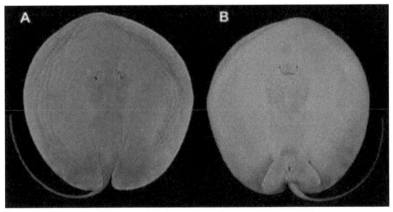

A. Dorsal view   B. Ventral view

Phylum: Chordata	Subphylum: Vertebrata
Class: Chondrichthyes	Subclass: Elasmobranchii
Order: Myliobtiformes	Family: Potamotrygonidae

Common name: Gomes's round ray.

Geographical distribution: Tropical; South America: Brazil in the upper Rio Amazonas basin and the lower Rio Amazonas basin.

Habitat: Freshwater; benthopelagic.

Distinctive features: Disk of this species has pancake-like appearance. It is also diagnosed by its unique dorsal color pattern which is composed of a uniform gray to light tan or brown color. Further, it has a slightly more slender tail width at base, small eyes and slightly greater preorbital snout length. A tiny, poisonous spine is seen on its tail. It grows to a maximum size of 13.5 cm DW (de Carvalho & Lovejoy, 2011).

Food habits: It feeds mainly on fish.

Reproduction: Reproductive mode is matrotrophic viviparity with tropho-nemata. Reproductive cycle is annual. Females of this species produce eggs, but these develop inside the female. Young hatch inside the female and are then born live.

Predators: Humans and caimans.

Parasites: Monogenea: *Potamotrygonocotyle tatianae.*

IUCN conservation status: Vulnerable.

5.1.26 HELIOTRYGON ROSAI (de Carvalho & Lovejoy, 2011)

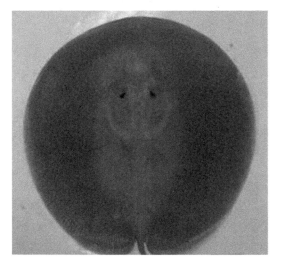

Common name: Rosa's round ray.

Geographical distribution: Tropical; South America: Brazil in the upper, mid, and lower Rio Amazonas basin and in the lower reaches of its major tributaries.

Habitat: Freshwater; benthopelagic.

Distinctive features: This species is identified by its unique dorsal color pattern which is composed of numerous white to creamy-white vermiculate markings over a light brown, tan, or gray background color. Its disk has a pancake-like appearance. Eyes are small. A tiny, poisonous spine is seen on its slender tail. This species grows to a maximum size of 80.0 cm DW (de Carvalho & Lovejoy, 2011).

Food habits: It feeds mainly on fish.

Reproduction: Reproductive mode is matrotrophic viviparity with trophonemata. Reproductive cycle is annual. Females of this species produce eggs, but these develop inside the female. Young hatch inside the female and are then born live.

Predators: Humans and caimans.

Parasites: Not reported.

IUCN conservation status: Not evaluated.

5.1.27 *PARATRYGON AIEREBA* (Müller & Henle, 1841)

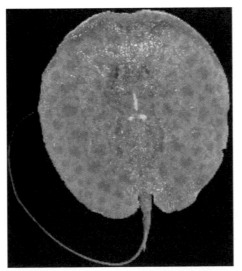

Phylum: Chordata   Subphylum: Vertebrata
Class: Chondrichthyes   Subclass: Elasmobranchii
Order: Myliobtiformes   Family: Potamotrygonidae

Common name: Discus ray.

Geographical distribution: Tropical; South America: Amazon River basin; Orinoco basin.

Habitat: Freshwater; demersal; common in shallow, sandy bottom areas, and near banks.

Distinctive features: Body of this species is exceptionally flat and oval, with light-brown coloration. Against this background, delicate, intricate designs with darker lines and dots appear. Tail is relatively thin. Maximum DW and weight of this species are 130 cm and 25 kg, respectively (http://akwa-mania.mud.pl/ryby/ryby/rybyp/Paratrygon aiereba. html).

Food habits: Diet of this species includes insects, crustaceans, and fishes.

Reproduction: This species is viviparous and has low fecundity, long gestation periods (9 months), and slow growth. It reproduces throughout the year and it may have one to eight intrauterine embryos. Females reach sexual maturity at 72 cm DW and males at 60 cm DW. Size at birth is 16 cm DW.

Predators: Not reported.

Parasites

Monogenea: *Potamotrygonocotyle aramasae*

Cestoda: *Anindobothrium guariticus*

Nandocestus copianullum

Nandocestus guariticus

Potamotrygonocestus fitzgeraldae

Potamotrygonocestus travassosi

Potamotrygonocestus sp.

Rhinebothrium brooksi

Rhinebothrium copianullum

Rhinebothrium sp.

IUCN conservation status: Data deficient.

5.1.28 *PLESIOTRYGON IWAMAE* (Rosa et al., 1987)

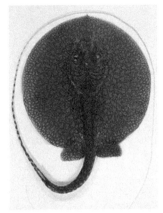

Phylum: Chordata   Subphylum: Vertebrata
Class: Chondrichthyes   Subclass: Elasmobranchii
Order: Myliobtiformes   Family: Potamotrygonidae

Common name: Long-tailed river stingray.

Geographical distribution: Tropical; South America: Upper to lower Amazon River basin, from Ecuador to Belém, Brazil, in the Napo, Solimões, Amazonas, and Pará rivers.

Habitat: Freshwater; demersal.

Distinctive features: This species has an oval disk which is longer than wide. Basic color of this species is pink to violet. Dorsal disk has brighter spots with small dots and belly is white. After the tail, color is white. Anterior disk is oval with a knob-like anterior protrusion. Eyes are very small and are not protruding from head. Spiracles are wide and rhomboidal and are much greater than reduced eyes. Teeth are set in quincunx, very small, and numerous (30–60/31–64 rows) with greater cusps in larger males. Usually, five buccal papillae are present inside mouth. Maximum DW and TL of this species are 65.0 and 200.5 cm, respectively (de Carvalho & Ragno, 2011).

Food habits: Feeds on small catfishes, insects, decapod crustaceans, molluscs, and parasitic cestodes and nematodes. It can detect electrical and chemical signals from prey in mud and sand (Rosa et al., 1987).

Reproduction: Gestation period of this species lasts up to 8 months and the time of birth occurs in the transition between the dry season and the wet season. Sexual maturity is attained by males when they reach 40 and

50 cm DW in the case of females. Litter size of this species ranges from 1 to 4 pups (average two pups).

Predators: Not reported.

Parasites

Cestoda: *Potamotrygonocestus chaoi*

Potamotrygonocestus marajoara

Copepoda: *Ergasilus trygonophilus*

IUCN conservation status: Data deficient; it is under the danger of extinction.

5.1.29 *PLESIOTRYGON NANA* (de Carvalho & Ragno, 2011)

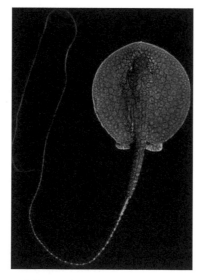

Common name: Dwarf river stingray.

Geographical distribution: Tropical; South America: Peru. Río Pachitea, tributary of Río Ucayali, up-river from town of Puerto Inca and Puerto Inca Province.

Habitat: Freshwater; demersal.

Distinctive features: Disk of this species is circular and as wide as long. Spiracles are faintly rhomboidal and very small ranging from 2.8% to 3.5% DW. Snout is very short. Eyes are very small that are directed with the lens upward and mouth is also very small. Tail fin length is about 130 cm in male and 150 cm in female. Dorsal color pattern of this species is composed of a dark-gray to dark-brown background color. Dorsal disk has also tan to yellow highly curved, slender and convoluted stripes or small spots forming rosette-like pattern, or with creamy white to yellowish irregularly shaped, scattered spots and ocelli. Disk diameter of males is 25–30 cm and in females, it is 30–35 cm (de Carvalho & Ragno, 2011; http://www.aquaristik-partner.de/galerie.html).

Food habits: Feeds on small catfishes, insects, decapod crustaceans, and molluscs.

Reproduction: This species is aplacental viviparous (ovoviviparous). Young are born with a diameter of 6–8 cm and a tail length of 40 cm. After 1 year, they can reach a diameter of 15 cm and a tail length of 80 cm. Sexual maturity occurs after 2–3 years with a diameter of 18–20 cm. A sexually matured female gives birth to 3–4 pups per year.

Predators: Not reported.

Parasites: Not reported.

IUCN conservation status: Least concern.

5.2   WHIPTAIL STINGRAYS (DASYATIDAE)

5.2.1   *MAKARARAJA CHINDWINENSIS* (Roberts, 2007)

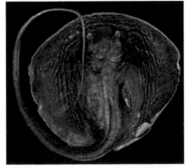

Phylum: Chordata	Subphylum: Vertebrata
Class: Chondrichthyes	Subclass: Elasmobranchii
Order: Myliobtiformes	Family: Dasyatidae

Common name: Not designated.

Geographical distribution: Tropical; Asia, Myanmar.

Habitat: Freshwater; benthopelagic.

Distinctive features: This species has a round disk. Dermal denticles are very minute. Floor of mouth is with four papillae: two in the middle and one to each side. Tail has a well-developed ventral cutaneous membrane. Sting insertion is relatively far posterior. Tail is long, with a low-lying fin fold. Base of the tail is more slender. Dorsal surface is pale gray or tan. Ventral disk surface is white. Tail is with distinct banding which is continued onto ventral cutaneous membrane. Spiral valve is thickened with 16–17 turns and 110 mm long. Reported values of maximum size and weight are 50 cm DW and 4 kg, respectively (Roberts, 2006).

No information is currently available on the habitat or biology of this species (Vishwanath, 2010).

IUCN conservation status: Data deficient.

5.2.2   *DASYATIS LAOSENSIS* (Roberts & Karnasuta, 1987)

Phylum: Chordata   Subphylum: Vertebrata
Class: Chondrichthyes   Subclass: Elasmobranchii
Order: Myliobtiformes   Family: Dasyatidae

Common name: Mekong stingray.

Geographical distribution: Asia: Chao Phraya and Mekong basins.

Habitat: Obligate freshwater species; prefers sandy substrate in large rivers; demersal.

Distinctive features: Disk is longer than wide. Pectoral fin disk of this species is oval in shape and slightly longer than wide. Tip of the snout protrudes. Eyes are small and followed by somewhat larger spiracles. There are 28–38 upper tooth rows and 33–41 lower tooth rows. While the teeth of juveniles and females are blunt, those of adult males are pointed with a central keel. A row of 5 papillae lie across the floor of the mouth. Five pairs of gill slits are short. Pelvic fins are longer than wide and triangular with rounded corners. Tail is whip like with 1–2 stinging spines on the upper surface. Both dorsal and ventral fin folds are found behind the spine. A single row of thorn-like dermal denticles runs along the midline of the back and tail with the largest found at the base of the tail. Number of enlarged midline denticles increases with age. Dorsal surface of the disk is brown. Middle disk is whitish with scattered large orange spots. Ventral surface of the disk is bright orange in color. Tail is whip-like and longer than body and caudal fin is absent. It grows to a maximum size of 62 cm DW and weight of 30 kg (Last & Compagno, 1999).

Food habits: Feeds on bottom-dwelling invertebrates.

Reproduction: This species exhibits ovoviparity (aplacental viviparity), with embryos feeding initially on yolk. Additional nourishment for the embryos is from the mother by indirect absorption of uterine fluid enriched with mucus, fat, or protein.

Predators: Not reported.

Parasites: Not reported.

IUCN conservation status: Critically endangered.

5.2.3   *DASYATIS GAROUAENSIS* (Stauch & Blanc, 1962)

Common name: Smooth freshwater stingray.

Geographical distribution: Tropical; Africa: Benue River in Cameroon and Nigeria; Niger river downstream; Lagos Lagoon and Cross river.

Habitat: Freshwater; demersal.

Distinctive features: A moderately large and thin-bodied ray in which disk is oval and flatter. Anterior margins of the disk are weakly concave. Denticles may be absent or restricted to central portion on dorsal surface of disk. Tip of snout projects as a small triangular process. Eyes are medium-sized and protruding, with small spiracles placed behind. Nostrils are covered by a flap of skin with a fringed posterior margin that reaches the small mouth. Teeth are small and closely set. There are 16–18 teeth in upper jaw and 14–28 teeth in lower jaw. Disk and pelvic fins are medium gray or gray-brown above and white below. Pelvic fins are roughly triangular. Tail which is darker or blackish and lighter below is whip-like and measures twice as long as the disk is wide. One or more stinging spines are positioned between one-sixth and one-fifth of the way along the tail. A narrow ventral fin fold originates closely behind the spine insertion. Coloration is plain brown or gray above, lightening at the disk margin, and white below with the fins outlined in starker white. Ampullae of Lorenzini of these species are smaller and simpler than those of marine stingrays. It grows to a maximum size of 40 cm DW (Compagno & Roberts, 1984).

Food habits: This species feeds on exclusively on aquatic nymphs of mayflies, stoneflies, caddisflies, and true flies.

Reproduction: This species exhibits ovoviparity (aplacental viviparity). Females of this species have a single functional ovary on the left side. Individuals mature at 2 years of age, with males living up to 5 years and females 7 years. Males have been reported to mature at 26.4 cm across and females at 26–30 cm across. Embryos feed initially on yolk, and subsequent nourishment from the mother by indirect absorption of uterine fluid enriched with mucus, fat, or protein through specialized structures.

Parasites: Not reported.

Predators: Not reported.

IUCN conservation status: Vulnerable.

5.2.4 *DASYATIS UKPAM* (Smith, 1863) = *Urogymnus ukpam*

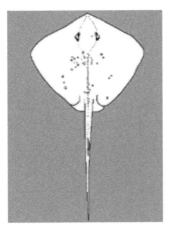

Common name: Thorny freshwater stingray.

Geographical distribution: Tropical; Africa: rivers and lakes in western Africa.

Habitat: Freshwater; brackish water; demersal; this obligate freshwater species is capable of tolerating higher salinities (euryhaline) and is able to move between different river systems through coastal waters.

Distinctive features: It is a very large and thick-bodied species. Entire dorsal surface of disk is covered with stout-spined denticles which are smooth in newborn. Sting may be greatly reduced in size or absent. When present, the spine averages 5.6 cm long in males and 4.6 cm long in female. It has a slightly projecting snout and an oval, very thick pectoral fin disk which is fairly longer than wide. Eyes are large and are followed by prominent spiracles. Mouth is slightly arched and contains many closely set, rounded teeth, numbering 38–40 rows in the upper jaw and 38–48 in the lower jaw. There are five papillae on the floor of the mouth. Pelvic fins are rounded and their inner margins are fused together. Tail is whip like with a narrow fin fold underneath and is becoming relatively shorter with age. Dorsal coloration is uniform dark brown or gray brown, and the tail is nearly black past the base. Belly is white with a broad dark edge around the margin of the disk. It grows to a maximum size of 120 cm DW and weight of 84.0 kg (Compagno, 1999).

Food habits: Its diet consists mainly of small eels.

Reproduction: This species exhibits ovoviparity (aplacental viviparity), with embryos feeding initially on yolk. Additional nourishment for the embryos is from the mother by indirect absorption of uterine fluid enriched with mucus, fat, or protein.

Predators: Not reported.

Parasites: Not reported.

IUCN conservation status: Endangered.

5.2.5   *HIMANTURA CHAOPHRAYA* (Monkolprasit & Roberts, 1990)

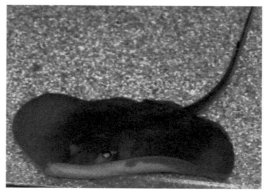

Phylum: Chordata   Subphylum: Vertebrata
Class: Chondrichthyes   Subclass: Elasmobranchii
Order: Myliobtiformes   Family: Dasyatidae

Common name: Giant freshwater stingray.

Geographical distribution: Most large rivers of tropical Australia and Fly River basin, New Guinea, the Mahakam River basin, Borneo, and several rivers in Thailand.

Habitat: Freshwater and brackishwater; bottom-dwelling species which favors a sandy or muddy habitat.

Distinctive features: This species has a thin, oval pectoral fin disk which is slightly longer than wide and broadest toward the front. Elongated snout has a wide base and a sharply pointed tip that projects beyond the disk. Eyes are minute and widely spaced. Large spiracles are behind the eyes. Mouth is small and forms a gentle arch and contains 4–7 papillae on the floor. Small and rounded teeth are arranged into pavement-like bands. There are five pairs of gill slits on the ventral side of the disk. Pelvic fins are small and thin. Matured males have fairly large claspers. The thin, cylindrical, and whip-like tail measures 1.8–2.5 times as long as the disk and lacks fin folds. A single serrated stinging spine is positioned on the upper surface of the tail near the base. There is band of heart-shaped tubercles on the upper surface of the disk extending from before the eyes to the base of the sting. There is also a midline row of four to six enlarged tubercles at the center of the disk. The remainder portion of the disk upper surface is covered by tiny granular denticles, and the tail is covered with sharp prickles past the sting. This species is plain grayish-brown above, often with a yellowish or pinkish tint toward the fin margins. Belly is white with broad dark bands, edged with small spots, on the trailing margins of the pectoral, and pelvic fins. Tail is black behind the spine. Maximum TL, DW and weight are 500 cm, 240 cm and 600 kg, respectively (Monkolprasit & Roberts, 1990).

Food habits: Diet consists of small fishes and invertebrates such as earth-worms, crustaceans, and molluscs, which it can detect using its electrore-ceptive ampullae of Lorenzini.

Reproduction: This species is aplacental viviparous, with the developing embryos nourished initially by yolk and later by histotroph (uterine milk) provided by the mother. Observed litter sizes range from 1 to 4 pups and the newborns measure around 30 cm across. Pregnant females are frequently found in estuaries, which may serve as nursery areas. Males mature sexually at approximately 1.1 m across and female maturation size and other life history details are unknown. Females have been reported to give birth to live young of 30 cm across.

Predators: Not reported.

Parasites (Fyler & Caira, 2006)

Cestodes: *Acanthobothrium asnihae*

Acanthobothrium saliki

Acanthobothrium zainali

Acanthobothrium etini

Acanthobothrium masnihae

Rhinebothrium abaiensis

Rhinebothrium kinabatanganensis

Rhinebothrium megacanthophallus

IUCN conservation status: Endangered overall, and critically endan-gered in Thailand.

5.2.6 *HIMANTURA DALYENSIS* (Last & Manjaji-Matsumoto, 2008)

Common name: Freshwater whipray

Geographical distribution: Northern Australia

Habitat: This demersal species is confined to freshwater and brackish water at a depth of 1–4 m. It is highly euryhaline and prefers a salinity range of 10–30 ppt.

Distinctive features: This species has an apple-shaped pectoral fin disk approximately as wide as long, with almost straight and transverse margins. Snout is moderately long and flattened with a pointed tip that projects from the disk. Eyes are small and are followed immediately by much larger spiracles. Mouth is gently arched and contains a pair of large papillae near the center and 2–3 much smaller papillae near the corners. There are about 37 upper tooth rows and 45 lower tooth rows and the teeth are small. Pelvic fins are small and the tail tapers from a narrow base to become thin and whip like and measure more or less twice the disk length. There is a single serrated stinging spine on top of the tail. Fin folds are absent. Entire upper surface of the disk is densely covered by small dermal denticles which are heart or oval-shaped around the shoulders and become tiny and granular toward the disk margins. There are also 5 relatively large denticles at the center of the disk. Tail is roughened by denticles above and below, with the largest positioned in a midline row before the spine.

Dorsal coloration is a uniform light brown to gray-brown, darkening to blackish past the tail spine. Belly and tail are white with dark-brown bands around the fin margins. This species attains a DW of 1.24 m.

Food habits: It feeds on small fishes and shrimps.

Reproduction: This species is aplacental viviparous, with the developing embryos nourished initially by yolk and later by histotroph (uterine milk) provided by the mother.

Predators: Not reported.

Parasites: Not reported.

IUCN conservation status: Vulnerable.

5.2.7 *HIMANTURA FLUVIATILIS* (Hamilton, 1822)

Common name: Ganges stingray.

Geographical distribution: Tropical; Asia: Gangetic river system; Bay of Bengal; and the Fly river system in New Guinea.

Habitat: Marine, freshwater, and brackishwater; demersal; also found in shallow bays and estuaries.

Distinctive features: Pectoral disk of this species is oval and slightly wider than long, with lateral margins rounded. Eyes are very small. Tail is whip-like, long, and less than twice the length of the disk. Entire dorsal surface, with the exception of the pelvic fins, is covered with dermal denticelli which are larger on the head. Some denticelli on the rear of the body are also rather large so as to resemble pointed thorns. Large areas of the abdomen are covered with tiny denticelli. Coloration is gray purple on the back, darker toward the edge of the disk, and lighter on the belly. Irregular dark bands extend along the lateral margins. This species is rather large and may reach a DW of at least 1.4 m and weight of 600 kg.

Food habits: It feeds on small fishes and shrimps, worms, and molluscs.

Reproduction: It exhibits ovoviparity (aplacental viviparity), with embryos feeding initially on yolk, and subsequently on additional nourishment from the mother by indirect absorption of uterine fluid (enriched with mucus, fat, or protein) through specialized structures. Fecundity of this species is very low (<10) (Talwar & Jhingran, 1991).

Predators: Not reported.

Parasites: Not reported.

IUCN conservation status: Endangered; it is threatened by habitat loss.

5.2.8   *HIMANTURA KITTIPONGI* (Vidthayanon & Roberts, 2005)

Common name: Roughback whipray.

Geographical distribution: Tropical; Asia: Thailand (mainstream of the Maekhlong river).

Habitat: Freshwater, demersal; prefers sandy or sandy–muddy bottom.

Distinctive features: It has a thin and oval pectoral fin disk which is slightly longer than wide, with the anterior margins converging at a broad angle to the tip of the snout. Snout terminates in a small, protruding knob. Eyes are small and are immediately followed by larger, teardrop-shaped spiracles. Central area of dorsal disk and tail has well-developed denticulation. A series of enlarged denticles is present on dorsal midline between pearl organ and base of sting. There are 4–5 papillae across the floor of the small mouth. Small teeth are arranged in 4–5 series in the upper jaw and 14–15 series in the lower jaw. Whip-like tail lacks fin folds and bears one or two serrated, stinging spines on the upper surface. It has a distinct coloration. Dorsal disk is with pale margin and pale spots are present in front of the eye and on the posterior rim of the spiracle. A narrow, dusky, gray, or brownish-orange marginal band is seen on the ventral disk surface. Belly is white. Tail is gray to orange-brown above and white below at the base, becoming white with dark spots or nearly black past the sting. It grows to a maximum size of 26.8 cm DW (Vidthayanon & Roberts, 2005).

Food habits: It feeds on small crustaceans and other benthic invertebrates.

Reproduction: Like other stingrays, it is aplacental viviparous with females providing their developing embryos with uterine milk. Males reach sexual maturity at around 25 cm across.

Predators: Not reported.

Parasites: Not reported.

IUCN conservation status: Endangered.

5.2.9 *HIMANTURA KREMPFI* (Chabanaud, 1923)

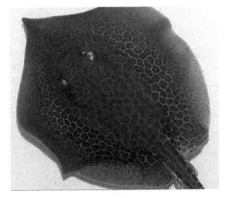

Common name: Marbled freshwater whip ray, leopard stingray.

Geographical distribution: Tropical; Asia: Chao Phraya (Thailand) and Mekong basins.

Habitat: Freshwater and brackishwater; demersal; estuaries and large rivers, often far upstream; prefers sandy substrates in which it can bury itself.

Distinctive features: This species has a thin, oval-shaped pectoral fin disk which is longer than wide. Snout is long and triangular with a pointed tip projecting from the disk. Eyes are small and are immediately followed by spiracles. Mouth is gently arched and contains an anterior row of four and posterior row of two papillae across the floor. There are 40–42 tooth rows in the upper jaw and 42–46 tooth rows in the lower jaw. Teeth are arranged with a quincunx pattern into pavement-like surfaces. Tail measures three times as long as the disk and bears two long stinging spines on top. After the spine, the tail becomes thin and whip like without any fin folds. There are many flattened, heart-shaped dermal denticles on the back which are arranged in a dense central band reaching the base of the tail and becoming smaller and sparser on the outer portions of the disk. Larger, heart-shaped denticles are also scattered over the disk, especially around the shoulders and the middle of the back. Two spines are present. Dorsal coloration is white to light gray, with brownish hexagonal blotches forming a reticulated pattern that fades toward the disk margin. Smaller individuals are covered by several irregular dark spots. Belly is entirely light colored. This species attains a maximum DW of 40 cm.

Food habits: It feeds on benthic invertebrates such as small crustaceans and molluscs.

Reproduction: It exhibits ovoviparity (aplacental viviparity), with embryos feeding initially on yolk and additional nourishment from the mother by indirect absorption of uterine fluid (enriched with mucus, fat, or protein) through specialized structures.

Predators: Not reported.

Parasites: Not reported.

IUCN conservation status: Endangered.

5.2.10 *HIMANTURA OXYRHYNCHA* (Sauvage, 1878)

Common name: Marbled whipray.

Geographical distribution: Tropical; Asia: Cambodia, Thailand, and Borneo.

Habitat: Freshwater and brackishwater estuaries; demersal; favors a sandy substrate in which it can bury itself.

Distinctive features: This species has a thin, oval-shaped pectoral fin disk which is longer than wide. Snout is long and triangular with a pointed tip projecting from the disk. Eyes are small and are immediately followed by spiracles. Mouth is gently arched and possesses an anterior row of four and posterior row of two papillae across the floor, which are followed by a seventh papilla in larger individuals. There are 40–42 tooth rows in the upper jaw and 42–46 tooth rows in the lower jaw. Teeth are arranged with a quincunx pattern into pavement-like surfaces. Tail measures three times as long as the disk and bears two long stinging spines on top. After the spine, the tail becomes thin and whip like without any fin folds. There are many flattened, heart-shaped dermal denticles on the back and are arranged in a dense central band reaching the base of the tail and becoming smaller and sparser on the outer portions of the disk. Larger, heart-shaped denticles are also scattered over the disk, especially around the shoulders and the middle of the back. Two spines are present. There is a line of 40–41 flat tubercles running down the dorsal midline, from between the eyes to the tail spines. Dorsal coloration is white to light gray, with brownish hexagonal blotches forming a reticulated pattern which fades toward the disk margin. Belly is entirely light colored. This species may attain a maximum DW of 36 cm.

Food habits: It feeds on benthic organisms such as small crustaceans and molluscs.

Reproduction: This species is ovoviviparous (aplacental viviparous) with females providing their unborn young with uterine milk as in other stingrays.

Predators: Not reported.

Parasites: Not reported.

IUCN conservation status: Endangered.

5.2.11 *HIMANTURA POLYLEPIS* (Bleeker, 1852)

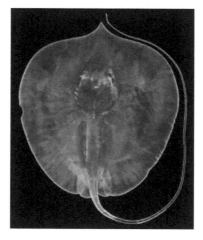

Common name: Giant freshwater stingray, freshwater whipray.

Geographical distribution: Tropical; Asia and Oceania: Mekong and Chao Phraya basins; eastern Borneo, New Guinea and northern Australia.

Habitat: Freshwater and brackishwater estuaries; demersal; potamodromous; prefers sandy bottoms.

Distinctive features: Dorsal surface of disk of this species has uniform brown or gray coloration. Caudal fin is absent. Tail is long and whip like. Ventral and dorsal skin folds are absent on the tail. A broad grayish to blackish marginal band is seen on the ventral surface of the disk. This species grows to a maximum DW, TL, and weight of 240 cm, 500 cm, and 600 kg, respectively (Rainboth, 1996).

Food habits: Feeds on benthic invertebrates and fishes.

Reproduction: It exhibits ovoviparity (aplacental viviparity), with embryos feeding initially on yolk and additional nourishment is from the mother by indirect absorption of uterine fluid enriched with mucus, fat, or protein through specialized structures. Young are born at about 30 cm DW.

Predators: Not reported.

Parasites: Not reported.

IUCN conservation status: Endangered.

5.2.12 *HIMANTURA SIGNIFER* (Compagno & Roberts, 1982)

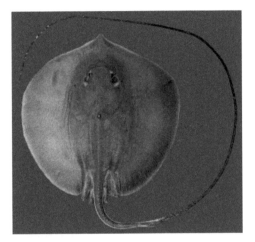

Common name: White-rimmed stingray.

Geographical distribution: Tropical; Asia: Indonesia, Malaysia, and Thailand (Chao Phraya, Mekong, and Tapi rivers).

Habitat: Freshwater and brackishwater estuaries; benthopelagic; prefers sandy bottoms.

Distinctive features: This species has a thin and oval pectoral fin disk which is slightly wider than long. Snout is slightly protruding. Eyes are small and are immediately followed by much larger spiracles. Mouth is gently bow shaped. There are 38–45 upper tooth rows and 37–46 lower tooth rows. Teeth are well spaced and arranged in a quincunx pattern. Each tooth has a blunt conical crown with a transverse cutting edge. There is a row of 4–5 papillae across the floor of the mouth. Tail is about 3.5 times as long as the disk and bears two stinging spines on top. After the spines, the tail becomes slender and whip like without any fin fold. Tail spines measure 86 mm long with about 70 serrations. Several small, flattened, heart-shaped dermal denticles interspersed with small conical denticles are present on the back. There are more denticles on the tail behind the spines also. This species has a narrow white marginal band around its disk. Disk is brown above, with mottling in the center and on the tail base. A white spot is seen anterior to spiracle and posterior to eye. Ventral surface is plain. It grows to a maximum size of 60 cm DW. TL may be greater than 200 cm (Kottelat et al., 1993).

Food habits: Feeds on bottom-dwelling crustaceans and shellfish.

Reproduction: This species breeds in freshwater and exhibits ovovi-viparity (aplacental viviparity), with embryos feeding initially on yolk. Additional nourishment for the embryos is from the mother by indirect absorption of uterine fluid (enriched with mucus, fat, or protein) through specialized structures. Newborns normally measure 10–12 cm across. Males and females mature sexually at 21–23 and 25–26 cm across, respectively.

Predators: Not reported.

Parasites: Not reported.

IUCN conservation status: Endangered.

KEYWORDS

- **freshwater stingrays**
- **common name**
- **distribution**
- **habitat**
- **species description**
- **food habits**
- **reproduction**
- **parasites**

CHAPTER 6

STINGRAY INJURIES, ENVENOMATION, AND MEDICAL MANAGEMENT

CONTENTS

ABSTRACT

Aspects relating to the venom glands and venom types; injuries, medical treatment, and prevention of injuries; and attacks of marine and freshwater stingrays are given in this chapter.

Among the different families of stingrays, Dasyatidae and Urolophidae comprise the most venomous stingrays with about 35 species under 4 genera. The dorsal location of the barbed tail spine in these stingrays makes these species more efficient stingers than other species. The longer the stingray spine and the more distally located the spine is on the whip-like tail, the greater the danger from stingray spine injuries. Interestingly, all the species of stingrays are venomous and are dangerous to humans. However, larger ones may inflict deeper and more painful wounds (https://answers.yahoo.com/question/index?qid=1006050233052).

6.1   STINGRAY'S VENOM APPARATUS

The stingray's venom apparatus is composed of the tail, or caudal appendage, along with a barbed spine and its enveloping integumentary sheath, and associated venom glands.

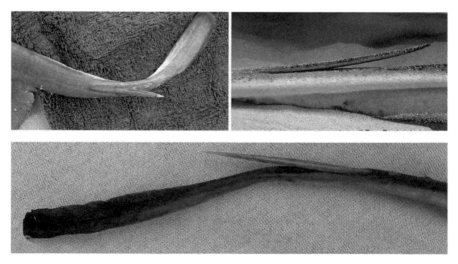

Spines in certain species of stingrays

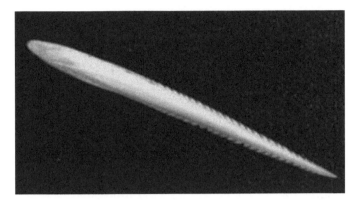

Tail spine showing barbs

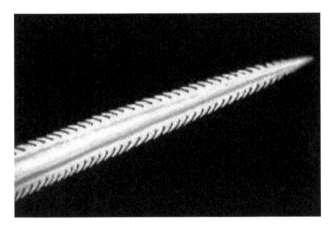

Magnified view of barbs

6.1.1 CHARACTERISTICS OF STINGER IN STINGRAYS

The stinger normally refers to the entire structure and its components, namely, the spine, its sheath, and the venom glands. The spine is the rigid surface of the sting, which is made of dentin. It is a razor-sharp, barbed, or serrated cartilaginous spine, which grows from the ray's whiplike tail and can grow as long as 37 cm. The barbs are the backward-facing serrations associated with the lateral aspect of the spine. Depending on the species, one to four stingers may be present on the dorsal surface of the tail. The barbs facilitate the tearing of the ray's integumentary sheath and the broadening of the victim's wound. The barbs also work like a backward-pointing fish hook. The stinger of the stingray is also a trauma- and venom-inducing apparatus. Its purpose may be purely defensive.

6.1.2 CHARACTERISTICS OF SPINES IN STINGRAYS

In the butterfly rays (Gymnuridae), the dorsal tail spines are less than 2.5-cm long and are attached to the root of the tail. In the eagle rays (Myliobatidae), the dorsal tail spines are up to 12 cm and are also proximally perched atop the root of the tail. In the dasyatid and urolophid stingrays, however, the dorsal tail spines are positioned more distally than in other species. The dasyatid stingrays, however, have longer spines (30 cm or longer in the larger species) than all other species of stingrays. The combination of distally placed long spines on long tails makes the dasyatid and urolophid stingrays the most dangerous group of stingrays. Humans are rarely killed unless the barb of the stingray punctures vital organs, such as the heart or lungs. Depending on the size of the stingray, the humans are usually stung in the foot region. Humans who harass stingrays have been known to be stung elsewhere, sometimes leading to fatalities.

6.2 STINGRAY VENOMS

The poisonous mucus (venom) from the barb of the giant freshwater stingray has been reported to share the characteristics with that of a Pit Viper with the poison inflicting a fast-acting necrosis and extreme pain to victims. The values of nitrogen, carbohydrates, and proteins in 100 mg of stingray venom are 3.1, 3.3, and 24.9 mg, respectively. The venoms of the *Potamotrygon scobina* and *Potamotrygon orbignyi* are known to induce edema (swelling) forming and nociceptive (nerve cell endings that initiate the sensation of pain) response in mice. The protein soluble and thermolabile toxin of the marine stingrays is composed of components such as serotonin, 5'-nucleotidase, and phosphodiesterase.

Serotonin: It causes severe vasoconstriction (painful severe contractions of smooth muscle). Further, it causes platelet activation ultimately leading to plug formation. In this case, the victim feels severe pain from the sting.

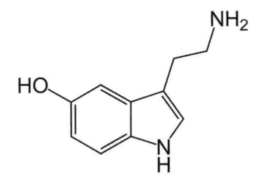

5′-Nucleotidase: It causes complete degradation of linear nucleic acids. Further, it causes necrosis (cell injury) and tissue breakdown (Hall et al., 2009).

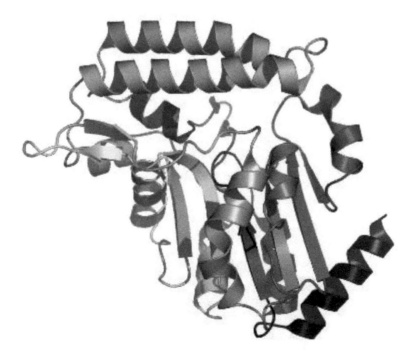

Phosphodiesterase: It causes hydrolysis of phosphodiester bonds in DNA and RNA. Further, it is responsible for the necrosis and tissue breakdown.

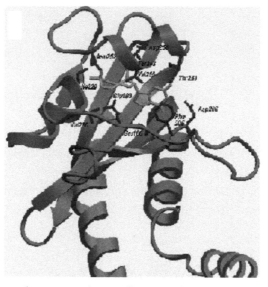

New toxins such as orpotrin, porflan, and hyaluronidase have also been recently isolated mainly from the freshwater stingrays. While the former two toxins are from *Potamotrygon orbigyni*, the last one is from *Potamotrygon motoro* (Conceição et al., 2006, 2009).

Orpotrin: It is a vasoconstrictor peptide causing arteriolar constriction.

Porflan: It is involved in the inflammatory processes.

Hyaluronidase: It causes hydrolysis of hyaluronic acid of extracellular matrix. Further, it has antigenic properties.

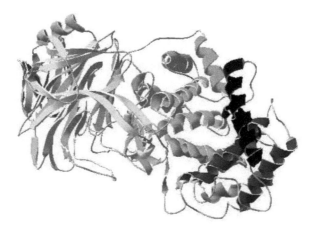

6.2.1 ENZYMATIC ACTIVITY OF STINGRAY VENOMS

In a study relating to the sting tissue extract of the marine stingray *Dasyatis guttata* and the fluvial stingray *Potamotrygon falkneri*, it was found that both the species contained proteolytic enzymes against casein, gelatin, and fibrinogen. While the crude *Potamotrygon henlei* venom showed hydrolytic action against gelatin, gelatinolytic activity was noticed in the venom of *Plesiotrygon iwamae*. On the other hand, both *P. scobina* and *P. orbygnyi* crude venoms showed low levels of proteolytic activity against casein. The proteolytic activity of venoms of stingrays against casein, gelatin, and fibrinogen is given in the following table (Ziegman & Alewood, 2015).

Species	Casein	Gelatin	Fibrinogen
Dasyatis guttata	+	+	+
Potamotrygon falkneri	+	+	+
Potamotrygon henlei		+	
Potamotrygon scobina	+		
Potamotrygon orbygnyi	+		
Plesiotrygon iwamae		+	

+, activity.

6.2.2 EDEMATIC AND NECROTIC ACTIVITIES OF STING VENOMS

The edema formation has been found to be dose dependent with the venom of the fluvial species *P. scobina* and *P. orbignyi*. It has also been found that incubating the venom at high temperatures attenuated the edema response. Further, *P. henlei* venom triggers a significant edematic response in mice paws. *P. motoro* venom, on the other hand, causes sustained edema for up to 48 h postinjection in mice. Interestingly, the spine extracts of *P. falkneri* produce not only mild inflammation in mice but also tissue necrosis as early as 3 h postinjection.

6.3 STINGRAY INJURIES

Several species of stingrays are usually very docile and they flee after any disturbance. However, certain larger species may be more aggressive and should only be approached with caution, especially in warm coastal areas. During the attack, the stingrays move their tail up and forward and stick their appendages into the body of the victims and liberate the poison (Rodríguez et al., 2008). Wounds associated with the stingray injuries have a traumatic (puncture) component and a toxic (envenomation) component. The puncture component is like a stiletto-type knife wound, most often inflicted on the lower leg of waders or the arm of fishermen. The danger of the venomous stingrays lies in the inoculation of the poison of the stings in the critical zones such as face, thorax, and abdomen. If a major blood vessel is lacerated during injury, the resulting hemorrhage may even be fatal. In some injuries, the spine gets broken off and remains in the wound. Introduction of the stingray's necrotizing venom directly into the body cavity of a person has been known to cause insidious necrotizing effects on the heart and other internal organs, and death is often inevitable (Rodríguez et al., 2008).

The stingray wounds often bleed profusely initially, followed by severe local pain progressing over the next 15–90 min. The direct trauma from the stinging barb is generally of much greater clinical significance than the envenoming syndrome. Affected victims may not be able to work for up to 8 months. Wounds associated with the swelling may also be infected with bacterial species such as *Streptococcus* spp., *Staphylococcus* spp., *Vibrio* spp., and *Mycobacterium marinum*.

6.3.1 INJURIES BY MARINE STINGRAYS

Among the marine stingrays, *D. guttata* is considered to be very dangerous. Marine injuries occur when the stingrays are accidentally captured in fishing nets or in shrimp trawling. Feet and ankles are the most affected organs of the people swimming in the shallow waters. Dangerous situations may also occur when these stingrays are captured in harpoons by divers. Deaths of swimmers are invariably due to heart and lung injuries provoked by stings of stingrays during the practice of swimming or snorkeling.

6.3.2 INJURIES BY FRESHWATER STINGRAYS

Among the freshwater stingrays, *P. falkneri* and *P. motoro* are considered to be very dangerous. *Potamotrygon* rays have robust tails with stingers in mid-distal position that frequently cause accidents. Victims are usually stung on the lower limbs during the day while swimming or moving through the water from a sand bar to another. Accidents usually occur in freshwater systems when riverine stingrays are stepped on or when cast nets are used. Accidents also occur in professional fishermen and river dwellers during the execution of services in ports or recreational activities along rivers (Haddad Junior et al., 2013).

6.3.3 SYMPTOMS ASSOCIATED WITH STINGRAY INJURIES

1. *Airways and lungs*: breathing difficulty, respiratory depression, and sweating;
2. *heart and blood*: irregular heartbeat and low blood pressure, syncope and hypotension, convulsions, arrhythmias (including bigeminy, all degrees of heart block and asystole), myocardial ischemia;
3. *nervous system*: fainting, generalized cramps, headache, paralysis, and weakness;
4. *muscular system*: muscle fasciculation and cramps;
5. *skin*: bleeding, pain, and swelling of lymph nodes near the area of the sting, severe pain at site of sting, sweating and swelling; and
6. *stomach and intestines*: diarrhea, nausea, and vomiting; cramping abdominal pain.

6.3.4 CLINICAL ASPECTS OF ENVENOMATION

Pain and tissue necrosis normally result at the sting site. The wounds are deep and wide and the torn tissue is painful, independent of the venom action. Initially, there are very strong cramps extending to the base of the injured limb. The injured limb is immobilized in flexion and painful contractures begin. Suffering becomes intolerable and the patient cannot stop screaming. The pain persists in the same intensity for 10 or 20 h and then slowly attenuates and only disappears a few days later. Shortly after the accident, the injured area begins to burn, turns hot and red, and the sensation is often accompanied by blood suffusion. When these wounds are poorly cared, lymphangitis (inflammation of the lymphatic system) and adenitis (inflammation of a gland) result. Later, the necrotic tissues are detached due to ulcers, causing spontaneous amputation of toes, feet, or hands. Healing always takes a long period.

6.3.5   LOCAL EFFECTS OF STINGRAY VENOM

1.   Severe pain at the site of injury; and
2.   Tissue necrosis—myonecrosis (in some situations involving cardiac muscle) is often noted in histological specimens of wounds.

6.3.6   DELAYED COMPLICATIONS OF STINGRAY INJURY

1.   Retained foreign bodies such as fragments of barbs, spine integument, and venom-secreting glandular tissue may migrate causing local tissue injury or result in granulomatous foreign body reactions;
2.   formation of pseudoaneurysms (leakage of arterial blood from an artery); and
3.   delayed infectious complications.

Local wound infection may be complicated by the following:

1.   Ecthyma gangrenosum (infection of the skin typically caused by *Pseudomonas aeruginosa*);
2.   osteomyelitis or septic arthritis (inflammation of the bone accompanied by bone destruction, usually due to bacterial infection);
3.   septicemia (also known as bacteremia or blood poisoning). It occurs when a bacterial infection enters the bloodstream;
4.   rarely necrotizing fasciitis (serious bacterial skin infection that spreads quickly and kills the body's soft tissue); and
5.   death due to tetanus.

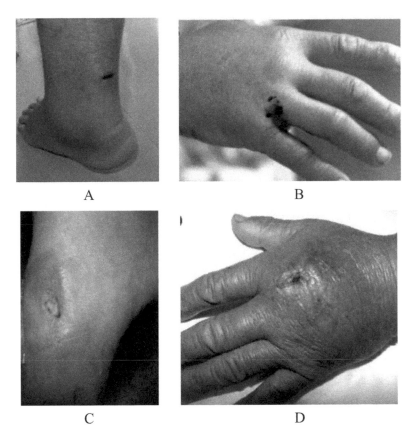

(A) Erythema and edema in a heel injury. (B) Hand injured by a stingray. (C) Ulcer on the left foot. (D) Hand stung.

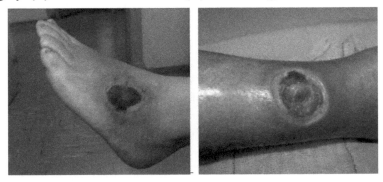

Skin necrosis and chronic ulcers in limbs

First aid and medical management of stingray trauma envenomation

6.4   MANAGEMENT OF STINGRAY INJURIES (JAMES, 2008)

The initial management of stingray injuries may begin at the scene immediately and be followed by wound exploration and management at nearby health-care facilities. The stingray-injured victims with thoraco-abdominal wounds and systemic manifestations should be referred to tertiary care facilities equipped and staffed for all imaging technologies (radiographs, ultrasound, and magnetic resonance imaging), critical care management, and cardiovascular surgery.

6.4.1   INITIAL MANAGEMENT

The initial management of stingray injuries should begin in the water as the victim is immediately assessed for cardiopulmonary stability. The wound is gently bathed in seawater to remove fragments of spine, glandular tissue, and integument. The victim should then be removed from the water. Any significant bleeding from lacerated vessels should be staunched with local pressure only. The wound should not be incised. The wound should be cleaned with freshwater or sterile irrigating solutions. Since stingray venom is heat labile, it is recommended for immediate immersion of the affected limb in warmed freshwater after immersing the uninjured contralateral limb first to assure safe water temperature and to prevent scalding.

6.4.2   TREATMENT

Treatment for stings of stingrays includes application of near-scalding water, which helps ease pain by denaturing the complex venom protein, and antibiotics. Immediate injection of local anesthetic in and around the wound is also very helpful. Local anesthetic brings almost instant relief for several hours. Any warm to hot fluid, including urine, has been reported to provide some relief. All stingray injuries should be medically assessed. The wound needs to be thoroughly cleaned, and surgical exploration is often required to remove any barb fragment remaining in the wound. Following cleaning, an ultrasound is helpful to confirm the removal of all the fragments. X-ray radiography imaging may be helpful where ultrasound is not available.

6.4.3 MEDICAL TREATMENT

Immediately after admitting the victim to a medical facility, he should be reassessed for cardiorespiratory stability. Tetanus prophylaxis should also be administered. If needed appropriate analgesia is done using parenteral analgesics. Then the wound should be carefully and thoroughly explored removing all fragments of spine, barbs, and foreign tissues. Radiographic examinations of the wound sites may reveal retained hyperdense, radio-opaque fragments of cartilaginous vasodentin spines, or barbs. Since these stingrays have cartilaginous endoskeletons, spine fragments may not be visible on conventional radiographic examinations. Magnetic resonance imaging and ultrasound examinations of wound sites may also help to locate hypodense, space-occupying retained foreign bodies, gas pockets, and cyst abscesses in septic wounds.

Expert surgical consultation may be obtained for the repair of the vascular injuries, lacerated nerves, and tendons, and for all stab wounds of the neck, thorax, or abdomen. Penetrating wounds of the neck, thorax, and abdomen may be associated with significant cardiovascular inju-ries including cardiac stab wounds and neurovascular bundle injuries. Such wounds will require surgical exploration under general anesthesia, removal of the stingray spine and its fragments, and repair of neurovas-cular and cardiac injuries. Cardiopulmonary bypass may be required for penetrating cardiac injuries. Pseudoaneurysm formation, and even arte-riovenous fistulae, may follow deeply penetrating stingray vascular inju-ries and require vascular repair of lacerated arteries and veins. Following wound exploration, debridement of devitalized tissues, and repair of vessel, nerve, or tendon injuries, stingray wounds may not be closed primarily but left open for drainage. Clean, superficial wounds may be loosely sutured and closed later. Deeply penetrating wounds may require additional surgical drainage and be allowed to remain open to heal. When stingray wound management is complicated by continuing tissue necrosis and ulceration, the topical application of recombinant human platelet-derived growth factor-BB (0.01% becaplermin gel) and hyperbaric oxygenation is used with standard wound management to successfully treat nonhealing stingray wounds (Diaz, 2008).

Antibiotics used in the treatment of uncomplicated infections and wound prophylaxis

1. Drug name: Levofloxacin (Levaquin).
    Adult dose 250–500 mg PO qd for 5 days.
    Pediatric dose <18 years: Not recommended.

2. Drug name: Cefixime (Suprax).
    Adult dose 400 mg/day.
    Pediatric dose <12 years: 8 mg/kg/day.

3. Drug name: Cephalexin (Keflex).
    Adult dose 250–1000 mg for 5 days.
    Pediatric dose 25–50 mg/kg/day.

4. Drug name: Doxycycline (Bio-Tab, Doryx, Vibramycin).
    Adult dose 100 mg for 5 days.
    Pediatric dose <8 years: Not recommended.

5. Drug name: Trimethoprim and sulfamethoxazole (TMP–SMZ, Bactrim, Septra).
    Adult dose 1 DS (double strength) tab for 5 days.
    Pediatric dose <2 months: Not recommended.

6.5   PREVENTION OF STINGRAY INJURIES AND ATTACKS (JAMES, 2008)

Stingray injuries normally occur in inexperienced and/or uniformed people grappling with live, terrified rays. Handling of aquarium captives must be kept to a minimum. Moving stingrays from one aquarium or transporting them should be done by devising some way of trapping them and then releasing them at their destination. All but the smallest stingrays should not be netted. Extreme caution must be exercised at all times. This might include the handler wearing gloves and a heavy long-sleeved shirt.

The most important factors in preventing infection are likely to be early wound irrigation, debridement, and the removal of any foreign material. Antibiotic cover should be considered for large or established wounds, especially if irrigation or debridement was delayed, and for penetrating wounds. Some more preventing measures are also given below:

1. to avoid the sea which is probably no guarantee of safety;
2. awareness of known stingray habitats;
3. to be extremely cautious if the water is murky;

4. to be wary of cruising along the sea bed when snorkeling or diving, particularly in known stingray habitats;
5. dive suits and dive boots offer little protection against a stingray barb;
6. to remember that the excised sting of a stingray may still be dangerous;
7. avoid handling a netted or hooked stingray; and
8. avoid trying to pet the friendly looking bull ray in your local aquarium.

KEYWORDS

- **stingers**
- **venom glands**
- **stingray injuries**
- **medicines**
- **treatment**

REFERENCES

Allen, G. R. *Field Guide to Marine Fishes of Tropical Australia*; Western Australian Museum—Science, 2009; p 256.

Babel, J. S. Reproduction, Life History, and Ecology of the Round Stingray, *Urolophus halleri* Cooper. *Fish Bulletin 137*, State of California—The Resources Agency, Department of Fish and Game Calisphere, University of California, 1967, http://content.cdlib. org/view?docId=kt6t1nb1vn&brand=calisphere&doc.view=entire_text.

Bauchot, M. L. Raies et autres batoides. In *Fiches FAO d'identificationpour les besoins de la pêche.* (rev. 1), Mèditerranée et mer Noire. Zone de pêche 37. Vol II; Fischer, W., Bauchot, M. L., Schneider, M., Eds.; Commission des Communautés Européennes and FAO: Rome, 1987; pp 845–886.

Benjamin, D.; Rozario, J. V.; Jose, D.; Kurup, B. M.; Harikrishnan, M. Morphometeric Characteristics of the Ornate Eagle Ray *Aetomylaeus vespertilio* (Bleeker, 1852) Caught Off Cochin, Southwest Coast of India. *Int. J. Environ. Sci.* **2012**, *3*, 685–688.

Bizzarro, J. J.; Smith, W. D. *Gymnura crebripunctata. The IUCN Red List of Threatened Species*, Version 2014.3, 2012.

Capapé, C.; Desoutter, M. Dasyatidae. In *Check-list of the Fishes of the Eastern Tropical Atlantic* (CLOFETA); Quero, J. C., Hureau, J. C., Karrer, C., Post, A., Saldanha, L. Eds.; JNICT/SEI/UNESCO: Lisbon/Paris/Paris, 1990; Vol 1, pp 59–63.

Charvet-Almeida, P.; de Almeida, M. P. *Potamotrygon magdalenae. The IUCN Red List of Threatened Species*, Version 2014.3, 2009.

Charvet-Almeida, P.; Soto, J. M. R.; de Almeida, M. P. *Potamotrygon brachyura. The IUCN Red List of Threatened Species*, Version 2014.3, 2009.

Chirichigno, N. F.; Vélez D. J. *Clave para identificar los peces marinos del Peru (Seguenda edición, revidada y actualizada)*; Instituto del Mar del Peru, Publicación especial, 1998; p 496.

Chisholm L. A.; Whittington, I. D. Two New Species of *Myliocotyle* (Monogenea: Monocotylidae) from the Gills of *Aetomylaeus maculatus and A. nichofii* (Elasmobranchii: Myliobatidae) from Sarawak, Borneo, Malaysia. *Folia Parasitol. (Praha)* **2004**, *51*, 304–310.

Compagno, L. J. V. Checklist of Living Elasmobranchs. In *Sharks, Skates, and Rays: The Biology of Elasmobranch Fishes*; Hamlett, W. C., Ed.; John Hopkins University Press, Baltimore, MD, 1999, pp 471–498.

Compagno, L. J. V.; Roberts, T. R. Dasyatidae. In *Checklist of the Freshwater Fishes of Africa (CLOFFA)*; Daget, J., Gosse, J. P., Thys van den Audenaerde, D. F. E., Eds.; ORSTOM/MRAC: Paris/Tervuren, 1984, Vol 1; pp 4–5.

Compagno, L. J. V. Dasyatidae. In *Smiths'Sea Fishes*; Smith, M. M.; Heemstra, P. C., Eds.; Springer-Verlag: Berlin, 1986; pp 135–142.

Comptes Rendus Biologies. *Comptes Rendus Biologies*, 2006.

Conceição, K.; Konno, K.; Melo, R. L.; Marques, E. E.; Hiruma-Lima, C. A.; Lima, C.; Richardson, M.; Pimenta, D. C.; Lopes-Ferreira, M. Orpotrin: A Novel Vasoconstrictor

Peptide from the Venom of the Brazilian Stingray Potamotrygon gr. orbignyi. *Peptides* **2006**, *27*, 3039–3046.

Conceição, K.; Monteiro-dos-Santos, J.; Seibert, C. S.; Silva, Jr., P. I.; Marques, E. E.; Richardson, M.; Lopes-Ferreira, M. *Potamotrygon* cf. *henlei* Stingray Mucus: Biochemical Features of a Novel Antimicrobial Protein. *Toxicon* **2012**, *60*, 821–829.

Conceição, K.; Santos, J. M.; Bruni, F. M.; Klitzke, C. F.; Marque, E. E.; Borges M. H.; Melo, R. L.; Fernandez, J. H.; Lopes-Ferreira, M. Characterization of a New Bioactive Peptide from Potamotrygon gr. orbignyi Freshwater Stingray Venom. *Peptides* **2009**, *30*, 2191–2199.

Cowley, P. D.; Compagno, L. J. V. A Taxonomic Re-evaluation of the Blue Stingray from Southern Africa (Myliobatiformes: Dasyatidae). *S. Afr. J. Mar. Sci.* **1993**, *13*, 135–149.

da Silva, J. P. C. B. M.; de Carvalho, M. R. A Taxonomic and Morphological Redescription of *Potamotrygon falkneri* Castex & Maciel, 1963 (Chondrichthyes: Myliobatiformes: Potamotrygonidae). *Neotrop. Ichthyol.* **2011**, *9*, 209–232.

de almeida, M. P.; R. B. Barthem, R. B.; da Silva, V. A.; Charvet-Almeida, P. Factors Affecting the Distribution and Abundance of Freshwater Stingrays (Chondrichthyes: Potamotrygonidae) at Marajó Island, Mouth of the Amazon River. *Pan-Am. J. Aquat. Sci.* **2009**, *4*, 1–11.

de Carvalho, M. R.; Lovejoy, N. R. Morphology and Phylogenetic Relationships of a Remarkable New Genus and Two New Species of Neotropical Freshwater Stingrays from the Amazon Basin (Chondrichthyes: Potamotrygonidae). *Zootaxa* **2011**, *2776*, 13–48.

de Carvalho, M.; Lovejoy, N.; Rosa, R. S. Potamotrygonidae (River stingrays). In Reis, R. E., Kullander, S. O., Ferraris, Jr., C. J., Eds. *Checklist of the Freshwater Fishes of South and Central America*; EDIPUCRS: Porto Alegre, Brasil, 2003; pp 22–28.

de Carvalho, M. R.; Ragno, M. P. An Unusual, Dwarf Species of Neotropical Freshwater Stingray, *Plesiotrygon nana* sp. nov., from the Upper and Mid Amazon Basin: The Second Species of *Plesiotrygon* (Chondrichthyes: Potamotrygonidae). *Papéis Avuls. Zool. (São Paulo)* **2011**, *51*, 101–138.

de Carvalho, M. R.; Sabaj Pérez, M. H.; Lovejoy, N. R. *Potamotrygon tigrina* a New Species of Freshwaters Stingray from the Upper Amazon Basin, Closely Related to *Potamotrygon schroederi* Fernandez-Yépez, 1958 (Chondrichthyes: Potamotrygonidae). *Zootaxa* **2011**, *2827*, 1–30.

Deynat, P. *Potamotrygon marinae* n. sp., a New Species of Freshwater Stingrays from French Guiana (Myliobatiformes, Potamotrygonidae). *Compt. Rend. Biol.* **2006**, *329*, 483–493.

Diaz, J. H. The Evaluation, Management, and Prevention of Stingray Injuries in Travelers. *J. Trav. Med.* **2008**, *15*, 102–109.

Dudley, S. F. J.; Kyne, P. M.; White, W. T. *Rhinoptera javanica. The IUCN Red List of Threatened Species*, Version 2014.3, 2006.

Duffy, C. *Myliobatis tenuicaudatus. The IUCN Red List of Threatened Species*, Version 2014.3, 2003.

Fahmi, W. T.; Manjaji, B. M.; Vidthayanon, C.; Badi, S.; Capuli, E. *Pastinachus solocirostris*. The IUCN Red List of Threatened Species. Version 2014.3, 2009.

Fontenelle, J. P.; Da Silva, J. P. C.; de Carvalho, M. R. *Potamotrygon limai*, sp. nov., a New Species of Freshwater Stingray from the Upper Madeira River System, Amazon Basin (Chondrichthyes: Potamotrygonidae). *Zootaxa* **2014**, *3765*, 249–268.

Fricke, R. Fishes of the Mascarene Islands (Réunion, Mauritius, Rodriguez): an annotated checklist, with descriptions of new species. *Theses Zoologicae*; Koeltz Scientific Books: Koenigstein, 1999, Vol 31; p 759.

Fyler, C. A.; Caira, J. N. Five New Species of *Acanthobothrium* (Tetraphyllidea: Onchobothriidae) from the Freshwater Stingray *Himantura chaophraya* (Batoidea: Dasyatidae) in Malaysian Borneo. *J. Parasitol.* **2006**, *92*, 105–125.

Góes de Araújo, M. L. *Potamotrygon schroederi. The IUCN Red List of Threatened Species*, Version 2014.3, 2009.

Góes de Araújo, P. C. M. L.; Pinto de Almeida, M. Reproductive Aspects of Freshwater Stingrays (Chondrichthyes: Potamotrygonidae) in the Brazilian Amazon Basin. *J. Northw. Atl. Fish. Sci.* **2005**, *35*, 165–171.

Grove, J. S.; Lavenberg, R. J. *The Fishes of the Galapagos Islands*; Stanford University Press, History, 1997; p 863.

Haddad Junior, V.; Cardoso, J. L. C.; Neto, D. G. Injuries by Marine and Freshwater Stingrays: History, Clinical Aspects of the Envenomations and Current Status of a Neglected Problem in Brazil. *J. Venom. Anim. Toxins Incl. Trop. Dis.* **2013**, *19*, 16. http://dx.doi.org/10.1186/1678-9199-19-16.

Hall, S.; Ishak, M. O.; Namdar, F; Hayes, H. Stingrays and the Toxicological Implications of their Defense Mechanism. *Biological Toxins* **2009**. http://www.timesonline.co.uk/tol/news/world/article627862.ece.

Heemstra, P. C.; Heemstra, E. *Coastal Fishes of Southern Africa—Marine fishes*. NISC (Pty.) Ltd., 2004, p 488.

Huveneers, C. *Dasyatis acutirostra. The IUCN Red List of Threatened Species*, Version 2014.3, 2006.

Ishihara, H.; Valenti, S. V. *Dasyatis ushiei. The IUCN Red List of Threatened Species*, Version 2014.3, 2009.

Isouchi, T. Butterfly Ray *Gymnura bimaculata*, a Junior Synonyms of *G. japonica. Jpn. J. Ichthyol.* **1977**, *23*, 242–244.

Izawa, K. Five New Species of *Eudactylina* Van Beneden, 1853 (Copepoda, Siphonostomatoida, Eudactylinidae) Parasitic on Japanese Elasmobranchs. *Crustaceana* **2011**, *84*, 1605–1634.

Jacobsen, I. *Gymnura tentaculata. The IUCN Red List of Threatened Species*, Version 2014.3, 2009.

James, H. D. The Evaluation, Management, and Prevention of Stingray Injuries in Travelers. *J. Trav. Med.* **2008**, *15*, 102–109.

Jeong, C. H.; Ishihara, H.; Wang, Y. *Myliobatis tobijei. The IUCN Red List of Threatened Species*, Version 2014.3.2008, 2009.

Kalidasan, K.; Ravi, V. M.; Sahu, S.; Maheshwaran, M. L.; Kandasamy, K. Antimicrobial and Anticoagulant Activities of the Spine of Stingray *Himantura imbricata. J. Coast. Life Med.* **2014**, *2*, 89–93.

Kirchhoff, K. N.; Klingelhofer, I.; Dahse, H. M.; Morlock, G.; Wilke, T. Maturity-related Changes in Venom Toxicity of the Freshwater Stingray *Potamotrygon leopoldi. Toxicon* **2014**, *92*, 97–101.

Klimley, A. P. *The Biology of Sharks and Rays*; University of Chicago Press—Science: Chicago, IL, 2013; p 524.

Koch, K. R.; Jensen, K.; Caira, J. N. Three New Genera and Six New Species of Leca-nicephalideans (Cestoda) from Eagle Rays of the Genus *Aetomylaeus* (Myliobati-formes: *M. myliobatidae*) from Northern Australia and Borneo. *J. Parasitol.* **2012,** *98*, 175–198.

Kottelat, M.; Whitten, A. J.; Kartikasari, S. N.; Wirjoatmodjo, S. *Freshwater fishes of Western Indonesia and Sulawesi*; Periplus Editions: Hong Kong, 1993; p 221.

Kyne, P. M.; Last, P. R. *Myliobatis hamlyni. The IUCN Red List of Threatened Species,* Version 2014.3, 2006.

Kyne, P. M.; Valenti, S. V. *Urobatis tumbesensis. The IUCN Red List of Threatened Species,* Version 2014.3, 2007.

Kyne, P. M.; Valenti, S. V. *Urotrygon venezuelae. The IUCN Red List of Threatened Species,* Version 2014.3, 2007.

Kyne, P.; White, W. T. *Urolophus circularis. The IUCN Red List of Threatened Species,* Version 2014.3, 2006.

Lamilla, J. *Urotrygon chilensis. The IUCN Red List of Threatened Species,* Version 2014.3, 2004.

Lamilla, J. *Myliobatis chilensis. The IUCN Red List of Threatened Species,* Version 2014.3, 2006a.

Lamilla, J. *Myliobatis peruvianus. The IUCN Red List of Threatened Species,* Version 2014.3, 2006b.

Lasso, C. A.; Rial B. A.; Lasso-Alcalá, O. Notes on the Biology of the Freshwater Stingrays *Paratrygon aiereba* (Müller & Henle, 1841) and *Potamotrygon orbignyi* (Castelnau, 1855) (Chondrichthyes: Potamotrygonidae) in the Venezuelan Llanos. *Aqua, J. Ichthyol. Aquat. Biol.* **1997,** *2*, 39–52.

Last, P. R.; Manjaji-Matsumoto, B. M. *Description of a New Stingray, Pastinachus gracilicaudus sp. nov. (Elasmobranchii: Myliobatiformes), Based on Material from the Indo-Malay Archipelago*, 2008. http://bionames.org/bionames-archive/issn/1833-2331/32/115.pdf.

Last, P. R.; Compagno, L. J. V. Dasyatididae. Stingrays. In *FAO Species Identification Guide for Fishery Purposes. The Living Marine Resources of the Western Central Pacific.* Vol. 3. *Batoid Fishes, Chimaeras and Bony Fishes Part 1 (Elopidae to Linophrynidae)*; Carpenter, K. E., Niem, V. H., Eds.; FAO: Rome 1999; pp 1479–1505.

Last, P. R.; Marshall, L. J. *Urolophus aurantiacus. The IUCN Red List of Threatened Species,* Version 2014.3, 2006a.

Last, P. R.; Marshall, L. J. *Urolophus javanicus. The IUCN Red List of Threatened Species,* Version 2014.3, 2006b.

Last, P. R.; Marshall, L. J. *Urolophus armatus. The IUCN Red List of Threatened Species,* Version 2014.3, 2006c.

Last, P. R.; Marshall, L. J. *Urolophus deforgesi. The IUCN Red List of Threatened Species,* Version 2014.3, 2006d.

Last, P.; Marshall, L. J. *Urolophus flavomosaicus. The IUCN Red List of Threatened Species,* Version 2014.3, 2006e.

Last, P. R.; Marshall, L. J. *Urolophus papilio. The IUCN Red List of Threatened Species,* Version 2014.3, 2006f.

Last, P. R.; Marshall, L. J. *Urolophus viridis. The IUCN Red List of Threatened Species,* Version 2014.3, 2006g.

Last, P. R.; Marshall, L. *Urolophus neocaledoniensis. The IUCN Red List of Threatened Species*, Version 2014.3, 2006h.

Last, P. R.; Marshall, L. J. *Trygonoptera galba. The IUCN Red List of Threatened Species*, Version 2014.3, 2009.

Last, P. R.; Manjaji-Matsumoto, B. M.; Kailola, P. J. *Himantura hortlei* n. sp., a New Species of Whipray (Myliobatiformes: Dasyatidae) from Irian Jaya, Indonesia. *Zootaxa* **2006,** *1239,* 19–34.

Last, P. R.; Manjaji-Matsumoto, B. M.; Moore, A. B. M. *Himantura randalli* sp. nov., a New Whipray (Myliobatoidea: Dasyatidae) from the Persian Gulf. *Zootaxa* **2012,** *3327,* 20–32.

Last, P. R.; Manjaji-Matsumoto, B. M.; Yearsley, G. K. *Pastinachus solocirostris* sp. nov., a New Species of Stingray (Elasmobranchii: Myliobatiformes) from the Indo-Malay Archipelago. *Zootaxa* **2005,** *1040,* 1–16.

Last, P. R.; Stevens, J. D. *Sharks and Rays of Australia*; CSIRO: Australia, 1994; p 513.

Loboda, T. S.; de Carvalho, M. R. Systematic Revision of the *Potamotrygon motoro* (Müller & Henle, 1841) Species Coplex in the Paraná-Paraguay Basin, with Description of Two New Ocellated Species (Chondrichthyes: Myliobatiformes: Potamotrygonidae). *Neotrop. Ichthyol.* **2013,** *11,* 693–737.

López M. I.; Bussing, W. A. *Urotrygon cimar*, a New Eastern Pacific Stingray (Pisces: Urolophidae). *Rev. Biol. Trop.* **1998,** *46* (Suppl.), 271–277.

Manjaji B. M.; Last, P. R. *Himantura lobistoma*, a New Whipray (Rajiformes: Dasyatidae) from Borneo, with Comments on the Status of *Dasyatis microphthamus. Ichthyol. Res.* **2006,** *53,* 290–297.

Manjaji, B. M.; Fahmi, W. T. *Himantura gerrardi. The IUCN Red List of Threatened Species*, Version 2014.3, 2009.

Manjaji, B. M.; Last, P. R.; Fahmi, W. T. *Himantura pastinacoides. The IUCN Red List of Threatened Species*, Version 2014.3, 2009.

Marshall, A. D.; Compagno, L. J. V.; Bennett, M. B. Redescription of the Genus *Manta* with Resurrection of *Manta alfredi* (Krefft, 1868) (Chondrichthyes; Myliobatoidei; Mobulidae). *Zootaxa* **2009,** *2301,* 1–28.

Masuda, H.; Amaoka, K.; Araga, C.; Uyeno, T.; Yoshino, T. *The Fishes of the Japanese Archipelago*; Tokai University Press: Tokyo, Japan, 1984; Vol. 1, p 437.

McEachran, J. D. Urolophidae. *Rayas redondas.* In *Guia FAO para Identificación de Especies para lo Fines de la Pesca. Pacifico Centro-Oriental*; Fischer, W., Krupp, F., Schneider, W., Sommer, C., Carpenter, K. E., Niem, V., Eds.; FAO: Rome, 1995; 3 Vols, pp 786–792.

McEachran, J. D.; di Sciara, G. N. Mobulidae. Mantas, diablos. In *Guia FAO para Identification de Especies para los Fines de la Pesca. Pacifico Centro-Oriental*; Fischer, W., Krupp, F., Schneider, W., Sommer, C., Carpenter, K. E., Niem, V., Eds.; FAO: Rome, 1995; 3 Vols, pp 759–764.

McEachran, J. D.; Séret, B. Gymnuridae. In *Checklist of the Fishes of the Eastern Tropical Atlantic (CLOFETA)*; Quero, J. C., Hureau, J. C., Karrer, C., Post, A.; Saldanha, L., Eds.; JNICT/SEI/UNESCO: Lisbon/Paris/Paris, 1990; Vol 1, pp 64–66.

Mejia-Falla, P. A.; Navia, A. F. New Records of *Urobatis tumbesensis* (Chirichigno & McEachran, 1979) in the Tropical Eastern Pacific. *Pan-Am. J. Aquat. Sci.* **2009,** *4,* 255–258.

Michael, S. W. *Reef Sharks and Rays of the World*; ProStar Publications—Science, 2005; p 107.

Miyake, T.; McEachran, J. D. Three New Species of the Stingray Genus *Urotrygon* (Myliobatiformes: Urolophidae) from the Eastern Pacific. *Bull. Mar. Sci.* **1988,** *42*, 366–375.

Monkolprasit, S.; Roberts, T. R. *Himantura chaophraya*, a New Giant Freshwater Stingray from Thailand. *Jpn. J. Ichthyol.* **1990,** *37*, 203–208.

Monteiro-dos-Santos, J.; Conceicao, K.; Seibert, C. S.; Marques, E. E.; Silva, Jr., P. I.; Soares, A. B.; Lima, C.; Lopes- Ferreira, M. Studies on Pharmacological Properties of Mucus and Sting Venom of *Potamotrygon* cf. *henlei*. *Int. Immunopharmacol.* **2011,** *11*, 1368–1377.

Moro, G.; Charvet, P.; Rosa, R. S. Insectivory in *Potamotrygon signata* (Chondrichthyes: Potamotrygonidae), an Endemic Freshwater Stingray from the Parnaíba River Basin, Northeastern Brazil. *Braz. J. Biol.* **2012,** *72*, 885–891.

Munro, I. S. R. *The Marine and Fresh Water Fishes of Ceylon*. Daya Books, Fishes, 2000; p 349.

Nishida, K. Phylogeny of the Suborder Myliobatidoidei. *Mem. Fac. Fish. Hokkaido Univ.* **1990,** *37*, 1–108.

Nishida, K.; Nakaya, K. *Taxonomy of the Genus Dasyatis (Elasmobranchia, Dasyatidae) from the North Pacific. NOAA Tech. Rept. NMFS* **1990,** *90*, 327–346.

Oldfield, R. G. Biology, Husbandry, and Reproduction of Freshwater Stingrays. II. *Trop. Fish Hobby* **2005a,** *54*, 110–112.

Oldfield, R. G. Biology, Husbandry, and Reproduction of Freshwater Stingrays. I. *Trop. Fish Hobby.* **2005b,** *53*, 114–116.

Rainboth, W. J. Fishes of the Cambodian Mekong. *FAO Species Identification Field Guide for Fishery Purposes*. FAO: Rome, 1996; p 265.

Ramírez, F.; Davenport, T. L. Ch. 2: Elasmobranchs from Marine and Freshwater Environments in Colombia: A Review. In *Colombia: Social, Economic and Environmental Issues*; Lavigne, G., Cote, C.; Nova Science Publishers: Hauppauge, NY, 2013; p 34.

Randall, J. E. *Coastal fishes of Oman*; University of Hawaii Press: Honolulu, HI, 1995; p 439.

Randall, J. E.; Allen, G.; Steene, R. C. *Fishes of the Great Barrier Reef and Coral Sea*; University of Hawaii Press: Honolulu, HI, 1990; p 506.

Rigby, C. *Himantura undulata. The IUCN Red List of Threatened Species*, Version 2014.3, 2012.

Roberts, T. R. *Makararaja chindwinensis*, a New Genus and Species of Freshwater Dasyatid Pastinachine Stingray from Upper Myanmar. *Nat. Hist. Bull. Siam Soc.* **2006,** *54*, 285–293.

Robertson, R.; Valenti, S. *Urotrygon simulatrix. The IUCN Red List of Threatened Species*, Version 2014, 2009a.

Robertson, R.; Valenti, S. V. *Urotrygon reticulata. The IUCN Red List of Threatened Species*, Version 2014.3, 2009b.

Robertson, R.; Rojas, R.; Valenti, S. V.; Cronin, E. *Urotrygon nana. The IUCN Red List of Threatened Species*, Version 2014.3, 2009.

Robins, C. R.; Ray, G. C. *A Field Guide to Atlantic Coast Fishes of North America*. Houghton Mifflin Company: Boston, MA, 1986; p 354.

Rodríguez, H. G. R.; Sánchez, E. C.; Méndez, J. D. Stingray Poisoning, A Careless Aspect in México. *Adv. Environ. Biol.* **2008,** *2*, 54–62.

Rosa, R.; Pinto de Almeida, M.; Charvet-Almeida, P. *Potamotrygon signata. The IUCN Red List of Threatened Species*, Version 2014.3, 2009.

Rosa, R. S.; Castello, H. P.; Thorson, T. B. *Plesiotrygon iwamae*, a New Genus and Species of Neotropical Freshwater Stingray (Chondrichthyes: Potamotrygonidae). *Copeia* **1987**, *2*, 447–458.

Rosa, R. S.; de Carvalho, M. R.; de Almeida Wanderley, C. *Potamotrygon boesemani* (Chondrichthyes: Myliobatiformes: Potamotrygonidae), a New Species of Neotropical Freshwater Stingray from Surinam. *Neotrop. Ichthyol.* **2008**, *6*, 1–8.

Ruocco, N. L.; Lucifora, L. O.; Díaz de Astarloa, J. M.; Delpiani., S. M. Morphology and DNA Barcoding Reveal a New Species of Eagle Ray from the Southwestern Atlantic: *Myliobatis ridens* sp. nov. (Chondrichthyes: Myliobatiformes: Myliobatidae). *Zool. Stud.* **2012**, *51*, 862–873.

Santos, H. R. S.; Gomes, U. L.; Charvet-Almeida, P. A New Species of Whiptail Stingray of the Genus *Dasyatis* Rafinesque, 1810 from the Southwestern Atlantic Ocean (Chondrichthyes: Myliobatiformes: Dasyatidae). *Zootaxa* **2004**, *492*, 1–12.

Schneider, W. *FAO Species Identification Sheets for Fishery Purposes. Field Guide to the Commercial Marine Resources of the Gulf of Guinea*; Prepared and Published with the Support of the FAO Regional Office for Africa: Rome, 1990; p 268.

Schwartz, F. J. Tail Spine Characteristics of Sting Rays (Order: Myliobatiformes) Frequenting the FAO Area 61 of the Northwest Pacific Ocean. *Raffles Bull. Zool.* **2007**, *14* (Suppl.), 121–130.

Séret, B.; Last, P. Description of Four New Stingarees of the Genus *Urolophus* (Batoidea: Urolophidae) from the Coral Sea, South-West Pacific. *Cybium* **2003**, *27*, 307–320.

Shen, A.; Ma, C.; Ni Y.; Xu, Z.; Ma, L. The Taxonomic Status of *Gymnura bimaculata* and *G. japonica*: Evidence from Mitochondrial DNA Sequences. *J. Life Sci.* **2012**, *6*, 9–13.

Smith, W. D.; Bizzarro, J. J. *Myliobatis longirostris. The IUCN Red List of Threatened Species*, Version 2014.3, 2006.

Stehmann, M.; McEachran, J. D.; Vergara R. Dasyatidae. In *FAO Species Identification Sheets for Fishery Purposes. Western Central Atlantic (Fishing Area 31)*; Fischer, W., Ed.; FAO: Rome, 1978, Vol. 1. [pag. var.].

Talwar, P. K.; Jhingran, A. G. *Inland Fishes of India and Adjacent Countries*. A. A. Balkema: Rotterdam, 1991; Vol 1, p 541.

The World's Marine Life. *The World's Marine Life*—National Fisheries Research Development Institute: Korea, 2009.

Thorson, T. B.; Langhammer, J. K.; Oetinger, M. I. Reproduction and Development of South American Freshwater Stingrays, *Potamotrygon mototro* and *P. cicularis. Environ. Biol. Fishes* **1983**, *9*, 3–24.

Treloar, M. A. *Urolophus expansus. The IUCN Red List of Threatened Species*, Version 2014.3, 2006.

Uyeno, T.; Matsuura, K.; Fujii, E., Eds. *Fishes Trawled Off Suriname and French Guiana, Japan*. Marine Fishery Resource Research Center: Tokyo, Japan, 1983; p 519.

Valenti, S. V. *Himantura marginata. The IUCN Red List of Threatened Species*, Version 2014.3, 2009a.

Valenti, S. V. *Potamotrygon marinae. The IUCN Red List of Threatened Species*, Version 2014.3, 2009b.

Valenti, S. V. *Urotrygon rogersi. The IUCN Red List of Threatened Species*, Version 2014.3, 2009c.

Valenti, S. V.; Kyne, P. M. *Pteromylaeus asperrimus. The IUCN Red List of Threatened Species*, Version 2014.3, 2009.

Valenti, S. V.; Robertson, R.; Vidthayanon, C.; Roberts, T. R. *Himantura kittipongi*, a New Species of Freshwater Whiptail Stingray from the Maekhlong River of Thailand (Elasmobranchii, Dasyatididae). *Nat. Hist. Bull. Siam Soc.* **2005**, *53*, 123–132.

Vishwanath, W. *Makararaja chindwinensis. The IUCN Red List of Threatened Species*, Version 2014.3, 2010.

Wang, Y.; Ebert, D.; Samiengo, B.; Vidthayanon, C. *Dasyatis navarrae. The IUCN Red List of Threatened Species*, Version 2014.3, 2009.

White, W. T. *Gymnura australis. The IUCN Red List of Threatened Species*, Version 2014.3, 2009.

White, W. T. *Aetobatus flagellum. The IUCN Red List of Threatened Species*, Version 2014.3, 2006a.

White, W. T. *Gymnura zonura. The IUCN Red List of Threatened Species*, Version 2014.3, 2006b.

White, W. T.; Furumitsu, K.; Yamaguchi, A. A New Species of Eagle Ray *Aetobatus narutobiei* from the Northwest Pacific: an Example of the Critical Role Taxonomy Plays in Fisheries and Ecological Sciences. *PLoS ONE* **2013**, *8*, e2013, *83785*, 1–12.

White, W. T.; Last, P. R.; Naylor, G. J. P.; Jensen, K.; Caira, J. N. Clarification of *Aetobatus ocellatus* (Kuhl, 1823) as a Valid Species, and a Comparison with *Aetobatus narinari* (Euphrasen, 1790) (Rajiformes: Myliobatidae). In *Descriptions of New Sharks and Rays from Borneo*; Last, P. R.; White, W. T.; Pogonoski, J. J., Eds.; CSIRO Marine and Atmospheric Research Paper No. 32, 2010; pp 141–164.

White, W. T.; Last, P. R.; Stevens, J. D.; Yearsley, G. K.; Fahmi, W. T. *Economically Important Sharks and Rays of Indonesia. [Hiu dan pari yang bernilai ekonomis penting di Indonesia]*. Australian Centre for International Agricultural Research: Canberra, Australia, 2006, http://dx.doi.org/.

White, W. T.; Manjaji, B. M.; Fahmi, W. T; Samiengo, B. *Himantura walga. The IUCN Red List of Threatened Species*, Version 2014.3, 2009.

White, W. T.; Moore, A. B. M. Redescription of *Aetobatus flagellum* (Bloch & Schneider, 1801), an Endangered Eagle Ray (Myliobatoidea: Myliobatidae) from the Indo-West Pacific. *Zootaxa* **2013**, *3752*, 199–213.

Yearsley, G. K.; Last, P. R. *Urolophus kapalensis* sp. nov., a New Stingree (Myliobatiformes: Urolophidae) Off Eastern Australia. *Zootaxa* **2006**, *1176*, 41–52.

Ziegman, R.; Alewood, P. Bioactive Components in Fish Venoms. *Toxins* **2015**, *7*, 1497–1531.

WEB REFERENCES

aquaportail.com.
de Araújo et al. http://www.cites.org/common/com/ac/20/E20-inf-08.pdf.
de Araújo et al., http://www.dfo-mpo.gc.ca/Library/315632.pdf.

de Araújo, M. L. G.; Charvet-Almeida, P.; Almeida, M. P.; Pereira, H. Freshwater Stingrays (Potamotrygonidae): Status, Conservation and Management Challenges, https://www.cites.org/common/com/ac/20/E20-inf-08.pdf.

de Araújo, M. L. G.; Charvet-Almeida, P.; Almeida, M. P.; Pereir, H. Freshwater Stingrays (Potamotrygonidae): Status, Conservation and Management Challenges, https://www.cites.org/common/com/ac/20/E20-inf-08.pdf.

de Carvalho, M. R.; Perez, M. H. S.; Lovejoy, N. R. *Potamotrygon tigrina*, a New Species of Freshwater Stingray from the Upper Amazon Basin, Closely Related to *Potamotrygon schroederi* Fernandez-Yepez, 1958 (Chondrichthyes: Potamotrygonidae). http://producao.usp.br/handle/BDPI/27625.

Encyclopaedia Animal 2—enciclopediaanimal@yahoo.com.ar.

FLMNH Ichthyology Dept., http://www.flmnh.ufl.edu/index.php?cID=1885.

FLMNH Ichthyology Dept., http://www.flmnh.ufl.edu/index.php?cID=2041.

FLMNH Ichthyology Dept., http://www.flmnh.ufl.edu/index.php?cID=2038.

FLMNH Ichthyology Dept., https://www.flmnh.ufl.edu/fish/discover/species-profiles/rhinoptera-bonasus.

FLMNH Ichthyology Dept., https://www.flmnh.ufl.edu/fish/discover/species-profiles/urobatis-jamaicensis/.

Florida Museum of Natural History—Ichthyology, http://www.flmnh.ufl.edu/index.php?cID=1859.

ftp://ftp.fao.org/docrep/fao/009/y4160e/y4160e39.pdf.

ftp://ftp.fao.org/docrep/fao/009/y4160e/y4160e40.pdf.

Góes de Araújo, M. L.; Charvet-Almeida, P.; Almeida, M. P.; Pereira, H. *Freshwater Stingrays (Potamotrygonidae): Status, Conservation and Management Challenges.* http://www.cites.org/common/com/ac/20/E20-inf-08.pdf.

http://akwa-mania.mud.pl/ryby/ryby/rybyp/Paratrygon aiereba.html.

http://animals.pawnation.com/bigtooth-stingrays-2890.html.

http://australianmuseum.net.au/common-stingaree-trygonoptera-testacea-muller-henle-1841.

http://biogeodb.stri.si.edu/sftep/en/thefishes/species/260.

http://fishdb.sinica.edu.tw/eng/species.php?id=383149.

http://media.utoronto.ca/media-releases/new-stingray-genus-discovered-in-the-amazon/.

http://rochen.chapso.de/potamotrygon-itaituba-s235039.html.

http://shark-references.com/literature/7570.

http://taibif.tw/zh/namecode/383163.

http://www.aquaristik-partner.de/galerie.html.

http://www.elasmodiver.com/River_Stingrays_Potamotrygonidae.htm.

https://www.flmnh.ufl.edu/fish/discover/species-profiles/pteroplatytrygon-violacea/.

http://www.heyaquarium.com/potamotrygon-leopoldi-freshwater-stingray-profiles/.

http://www.heyaquarium.com/potamotrygon-sp-itaituba-profile-freshwater-ray/.

http://www.mexfish.com/fish/lneagle/lneagle.htm.

http://www.mexfish.com/mexico/panamic-stingray/.

http://www.mexfish.com/mexico/spiny-stingray/.

http://www.producao.usp.br/handle/BDPI/27625.

http://www.raylady.com/Potamotrygon/species/Yepezi.html.

http://www.researchgate.net/profile/Joao_Paulo_Da_Silva4/publication/270760341_
    Familia_Potamotrygonidae_-_Potamotrygon_humerosa/links/54b41b880cf28ebe92e
    4535b.pdf.

http://www.timesonline.co.uk/tol/news/world/article627862.ece.

http://zukan.com/fish/internal842.

https://answers.yahoo.com/question/index?qid=1006050233052.

https://rybicky.net/atlasryb/trnucha_castexova.

Last, P. R.; Manjaji-Matsumoto, B. M. Description of a New Stingray, *Pastinachus gracili-
    caudus* sp. nov. (Elasmobranchii: Myliobatiformes), Based on Material from the Indo–
    Malay Archipelago. http://bionames.org/bionames-archive/issn/1833-2331/32/115.pdf.

Last, P. R.; Fahmi; Naylor, G. J. P. *Pastinachus stellurostris* sp. nov., a New Stingray (Elas-
    mobranchii: Myliobatiformes) from Indonesian Borneo. http://bionames.org/bionames-
    archive/issn/1833-2331/32/129.pdf.

Last, P. R.; White, W. T. http://bionames.org/bionames-archive/issn/1833-2331/22/275.pdf

Lizard Island Field Guide, http://lifg.australianmuseum.net.au/Group.html?groupId=eOb
    QN8ee&hierarchyId=PVWrQCLG.

Martin, R. A.; MacKinlay, D. Biology and Conservation of Freshwater Elasmobranchs. In:
    Symposium Proceedings, International Congress on the Biology of Fish Tropical Hotel
    Resort, Manaus Brazil, August 1–5, 2004, http://www.dfo-mpo.gc.ca/Library/315632.pdf.

Mexico Fish, Flora, & Fauna, http://www.mexican-fish.com/longnose-eagle-ray/.

Mexico Fish, Flora, & Fauna, http://www.mexfish.com/mexico/panamic-stingray/

Mexico Fish, Flora, & Fauna, http://www.mexfish.com/mexico/spiny-stingray/

River Stingrays, The Elasmodiver Shark and Ray Field Guide, http://www.elasmodiver.
    com/River_Stingrays_Potamotrygonidae.htm.

www.imms.org.

INDEX